Penguin Education
Studies in Applied Statistics
General Editor: I. R. Vesselo

Statistics for the Social Scientist: 1
Introducing Statistics

K. A. Yeomans

Statistics for the Social Scientist: 1
Introducing Statistics

K. A. Yeomans

Penguin Education

Penguin Education,
A Division of Penguin Books Ltd,
Harmondsworth, Middlesex, England
Penguin Books Inc, 7110 Ambassador Road,
Baltimore, Maryland 21207, U.S.A.
Penguin Books Australia Ltd, Ringwood,
Victoria, Australia

First published 1968
Reprinted 1970, 1971, 1973
Copyright © K. A. Yeomans, 1968

Made and printed in Great Britain by
Western Printing Services Ltd, Bristol
Set in Monotype Times

To my parents

Contents

8   Contents

Editorial foreword

The rapid increase in the use of statistical techniques, especially in experimental fields, has made necessary a wider understanding both of the foundations of these techniques and of their limitations.

This series of *Studies in Applied Statistics* has been produced to meet the needs of the student or worker in a special field of interest who needs to be informed of up-to-date statistical methods and how to use them, but who will not be content to accept the results of others and the rules of calculation without rather more explanation than is usual. Although not necessarily able to derive the formulae personally, he will wish at least to satisfy himself of their plausibility and to have some confidence in his interpretation of the results of calculation.

Statistics for the Social Scientist: 1 Introducing Statistics deals with the methods, numerical and graphical, for describing a set of statistical data; that is, presenting them in a form which is readily intelligible to the layman. Since there are so many students and workers whose knowledge of the rudimentary processes of mathematics is either restricted or forgotten, a very early start is made. It is hoped that, since these problems are common to many fields, volume one may also be useful as a background in other studies.

Volume 2 continues the work through the field of applied statistics, with special reference to the areas most useful to the social scientist.

The author is a young man who has shown a deep understanding of the special difficulties of his non-mathematical students, and he has brought this insight to bear on his work.

The general editor would like to acknowledge the valuable help of Mr Arthur Tulip, Miss Angela Powell and Miss Jacqueline Smith, all of Honeywell Controls Ltd, in connexion with the computation of tables.

I. R. V.

Preface

Whilst teaching statistical method to social science students I have become aware of certain deficiencies in the range of available textbooks. By and large these students have a limited mathematical background which makes mathematical statistics textbooks inappropriate, but on the other hand dissatisfaction is generated by the 'take it or leave it' approach of many introductory texts which produce formulae and methods like rabbits out of a hat. A compromise between these two extremes seemed necessary.

Introducing Statistics is aimed at the non-mathematician and starts with a revision of simple but important mathematical ideas. Wherever possible a mathematical treatment has been avoided in the text, although mathematical notes occur at the end of chapters for readers wanting a more rigorous approach. The student should find that the scope of this volume will provide the essential material for a one session course. If it is followed by *Statistics for the Social Scientist: 2 Applied Statistics*, where recent developments and applications are considered at a more advanced level, a sound grounding in the techniques and methods of statistical analysis should have been obtained.

The word 'statistics' conjures up different impressions to different people. It is generally recognized that figures (numerical data) are collected, arranged and presented in various forms by the statistician. It is also a commonly accepted belief that anything can be proved or disproved with these statistics. There is some truth in these observations, but both fall short of the true purpose and approach of the subject.

Certainly the collection, arrangement and presentation of data are of importance, but so is the analysis and interpretation of the data collected. Statistical method enables empirical evidence to be used for assessing previously hypothesized situations and relationships and it is true that inappropriate use of the available tools can produce misleading conclusions, but misuse is not confined to statistics. On the credit side, the ability to describe and compare phenomena can be of great benefit to the policy maker or the decision taker. There are many defects in our society just as there are worthwhile aspects of the social, economic, educational, industrial and administrative framework, but before ills can be remedied or success perpetuated we must establish the facts with precise information based upon sound methods.

It is possible to view the tools and methods of statistics in isolation using a generalized theoretical approach which can have relevance for any subject ('mathematical statistics'). From this field come the precise methodology and the development of new techniques, but side by side with 'pure' statistics we must recognize the applications of the tools developed. This will particularly interest the social scientist to whom the statistical method is useful only if it is an aid in solving problems; for him the elements of theory need not be rigorously established, demonstration by illustration and example being sufficient to enable the principles and their applications in the social sciences to be developed simultaneously.

K. A. Yeomans
January 1968

Acknowledgements

I should like to express my indebtedness and gratitude to the many people who have given their advice and encouragement unstintingly. At both the Lanchester College of Technology and at the Birmingham College of Commerce my colleagues were always ready to listen to and discuss difficulties. To mention a few is an injustice to the rest. Nevertheless, I feel that Mr R. B. Goudie of the Birmingham College of Commerce and Dr C. Sharp of the University of Leicester should be given my special thanks. Mr Goudie has read the majority of the book at manuscript stage and has provided considerable technical and administrative assistance throughout its production. Many mistakes and ambiguities would have gone unnoticed but for his vigilance. Dr Sharp also kindly read much of the manuscript and the majority of his valuable suggestions have been incorporated in the following chapters. Last but no means least I thank my wife, Merle, who has so often provided the encouragement which has lifted my flagging spirits, and who has somehow managed to fit into the most hectic day of looking after a young and active family the typing of the manuscript.

I would also like to acknowledge permission given to publish the following copyright material:

(i) Her Majesty's Stationery Office for permission to use Crown copyright material taken from the *Annual Abstract of Statistics, National Income and Expenditure, Monthly Digest of Statistics, U.K. Balance of Payments* and *Regional Statistics*.
(ii) The Universities of Birmingham, Bristol and London for permission to use questions from their examination papers; I would point out that

these bodies are in no way committed to approval of the answers given in this book.

(iii) Barclays Bank for permission to use series taken from *Barclays Bank Review*.

K. A. Yeomans

Chapter 1
The mathematical basis of statistics

The purpose of this chapter is to consider the tools of mathematics which are used in the study and application of statistics. The rules and principles are very simple and their inclusion may seem superfluous, but they are worth stating explicitly for revision purposes. Many students of the human sciences will have to face a course in statistics after several years with little or no contact with mathematics and there is a need to refresh their memories, to reconsider the fundamental rules, and to extend their knowledge to cover the arithmetical and algebraic principles specifically required in statistics.

1.1 Shorthand and definitions

The symbols $+$, $-$, $\div$ and $\times$, indicating addition, subtraction, division and multiplication, probably need little discussion. It is necessary, however, to mention that brackets can represent multiplication, as can a dot, or two symbols side by side. Thus 3×8 can be written $(3)(8)$, and $x \times y$ can be shown by $x.y$ or simply xy. Similarly the division sign is not always used; frequently $5 \div 8$ will be shown by $5/8$ or $\frac{5}{8}$.

1.1.1 Other symbols which will be encountered are $>$, $<$, $\geqslant$, and $\leqslant$. If $r > 0.8$ we mean that r is greater than 0.8; by $r < 0.8$ we mean that r is less than 0.8. In both cases the open end of the symbol shows the value which is larger. $\geqslant$ and $\leqslant$ have similar meanings except that for $r \geqslant 0.8$ and $r \leqslant 0.8$ we are now allowing r to be equal to 0.8, whereas this was previously denied.

This introduces us to the equals sign, $=$. Its use is fairly evident in the context of ordinary arithmetic, e.g. $9 - 3 = 6$, but modifications are often found, such as $\simeq$, $\doteqdot$ or $\simeq$ to indicate that something is approximately equal to something else (for instance, $18 \cdot 95 \times 4 \cdot 32 \simeq 82$). The sign $\neq$ shows that two values are not equal, for example $\dfrac{15+2}{3} \neq 15+\tfrac{2}{3}$ and an extension of the equals sign is $\equiv$ which means that two things are identical, i.e. $(x+y)^2 \equiv x^2+2xy+y^2$.

1.1.2   Words such as product, dividend, numerator and so on are often used by teachers of mathematics and statistics. These are sometimes poorly understood by the student, so let us define them specifically.

In $5 \times 18 = 90$, 5 is the *multiplicand*, 18 is the *multiplier* and 90 is the *product*. For $24 \div 6 = 4$, 24 is the *dividend*, 6 is the *divisor* and 4 is the *quotient*. Finally, in the fraction $\tfrac{8}{9}$, 8 is the *numerator* and 9 is the *denominator*.

1.2 Procedure and method in arithmetic

1.2.1 *Mixed calculations*

Where a problem involves addition and subtraction together with multiplication and division, the multiplication and division should be performed first. As examples consider $8-3 \times \tfrac{6}{9} = 2$, or $7+6 \times 4-2 \times 3 = 25$. The only exception to this rule will be where certain figures are enclosed in brackets, so that $(8+5) \times (6+1) = 13 \times 7 = 91$, and $\dfrac{(3+2) \times 2}{3(6+1)} = \dfrac{10}{21}$. In all such cases the terms within the bracket should be treated as a single number, and therefore must be resolved first. If the bracket is to be removed and the sign preceding it is negative, then all the signs inside are reversed. Therefore $(9+1-4)-(6+2-3) = 9+1-4-6-2+3 = 1$.

1.2.2 *Fractions and decimals*

The numerator and denominator of a fraction can be multiplied or divided by the same number or symbol without altering the real meaning of the fraction. $\dfrac{4}{5}$ is the same as $\dfrac{4 \times 6}{5 \times 6} = \dfrac{24}{30}$, and $\dfrac{10}{35}$ is the same as $\dfrac{\frac{10}{5}}{\frac{35}{5}}$. The same number or symbol cannot be added to or subtracted from the numerator and denominator of a fraction without changing the fraction's value: $\dfrac{5}{11}$ is not the same as $\dfrac{5-4}{11-4}=\dfrac{1}{7}$. Fractions can only be added or subtracted after their denominators have been transformed to a common form. Thus $\dfrac{8}{9}+\dfrac{2}{3}=\dfrac{8}{9}+\dfrac{6}{9}=\dfrac{14}{9}$. The general rule is to find a new denominator (as small as possible) into which each original denominator can be divided exactly. The resulting quotient in each case is multiplied by the original numerator and the two values added or subtracted giving a new numerator. The above illustration could have been resolved in this way: $\dfrac{8}{9}+\dfrac{2}{3}=\dfrac{8+6}{9}=\dfrac{14}{9}$. Multiplication and division involving fractions is very much simpler. For multiplication the product of the numerators over the product of the denominators provides the correct answer $\left(\dfrac{1}{6}\times\dfrac{3}{7}=\dfrac{3}{42}\right)$ while for division the divisor fraction is inverted and multiplied by the dividend fraction:

$$\left(\dfrac{3}{6}\div\dfrac{6}{4}=\dfrac{3}{6}\times\dfrac{4}{6}=\dfrac{12}{36}\right)$$

Common fractions have different denominators. Often it is preferable to reduce them to a common basis of 10 or some power of 10 to give decimal fractions. This is achieved by dividing the denominator of the fraction into the numerator. As examples consider $\dfrac{2}{5}=\dfrac{4}{10}=0\cdot4$; $\dfrac{18}{50}=\dfrac{36}{100}=0\cdot36$; $\dfrac{1}{90}=\dfrac{111\frac{1}{9}}{10000}=0\cdot0111'$ (the dash indicates that the 1 recurs continuously). The principle is self-evident. If the numerator is smaller than the denominator, add a zero to the numerator (increase it by a multiple of 10). The first figure in the answer is then preceded by the point. If it is necessary to increase the numerator by 100, then the first figure of the quotient will be preceded by the point and one zero. This can be continued as necessary. In a similar way, decimals can be converted to fractions by putting their absolute value over 10, 100, 1000 etc., whichever is appropriate. For instance,

$$8\cdot2=\dfrac{82}{10};\ 0\cdot65=\dfrac{65}{100};\ 0\cdot019=\dfrac{19}{1000}$$

The denominator will be a figure 1 followed by as many zeroes as there are figures (including zeroes) after the decimal point. Addition and subtraction are straightforward, except that care must be taken to ensure that the decimal points are lined up. In multiplication, the product has as many decimal places as there are in the multiplicand and multiplier together. Therefore $0{\cdot}93 \times 2 = 1{\cdot}86,\quad 0{\cdot}003 \times 0{\cdot}001 = 0{\cdot}000003,$ $1{\cdot}8 \times 2{\cdot}5 = 4{\cdot}50.$

The best way to carry out division of decimals is to multiply both numerator and denominator by some power of 10 so as to convert the denominator into a number between 1 and 10 (i.e. with the decimal point after the first non-zero digit). Then the division proceeds easily as in the following examples:

$$\frac{0{\cdot}036}{0{\cdot}24} = \frac{0{\cdot}36}{2{\cdot}4} = 0{\cdot}15$$

$$\frac{0{\cdot}036}{0{\cdot}48} = \frac{0{\cdot}36}{4{\cdot}8} = 0{\cdot}075$$

$$\frac{0{\cdot}01}{0{\cdot}0025} = \frac{10}{2{\cdot}5} = 4{\cdot}0$$

1.2.3 *Positive and negative numbers*

Addition of numbers each with the same sign involves simple summation and the adoption of the common sign. Addition of numbers with unlike signs involves summation of the negative and positive values separately, and finally the subtraction of the smaller absolute figure from the larger. The answer takes the sign of the larger figure. Therefore:

$$(+8)+(+6)+(-3)+(-6) = 14-9 = +5$$
$$(+2)+(-10)+(-9)+(+7) = -19+9 = -10$$

Subtraction of one signed number from another involves changing the sign of the second and adding the result to the first:

$$8-(-3) = 11; \; -10-(+6) = -16$$

This agrees with the rules for brackets laid down earlier. Multiplication and division of two numbers with like signs will always give a positive answer, while the presence of unlike signs will always mean a negative answer. Thus $-8 \div 2 = -4$, and $-8 \times -6 = +48$.

1.2.4 *Zero*

The addition or subtraction of 0 to or from any number or symbol leaves the number or symbol unaltered. Zero multiplied by any number or symbol always equals 0. Zero divided by a number or symbol gives a quotient of 0, while the use of 0 as a divisor is not permitted.

1.3   Simple algebra

We have seen in passing that the rules of arithmetic applying to numerical values equally apply to algebraic symbols. Let us examine one of the fundamental expressions in algebra, namely the equation, and specify how we should deal with it.

1.3.1   *Simple equations*

The simplest type of equation is $5x = 15$. It tells us something about the unknown value of x. Rearrangement enables us to discover that x is equated to $\dfrac{15}{5} = 3$. There are of course many other forms of equation which may be much more complicated. They may contain variables, usually shown by the letters x and y, and also constants, shown by the first letters of the alphabet. Each family of equations has its own set of characteristics which are perhaps best shown graphically.

The linear equation in the form $y = a+bx$ is the one which we shall encounter most often in statistics. If we know a and b we shall find that as x changes so does y (this is why x and y are called variables). We should notice particularly that the amount by which y changes as the result of a given change in x is determined by the values and signs of a and b. If we have the following linear equation, $y = 2+0.5x$, and we substitute different values of x, we will find corresponding values of y as follows:

$$\text{for } y = 2+0.5x, \text{ when } x = -3, y = 0.5$$
$$x = -2, y = 1.0$$
$$x = -1, y = 1.5$$
$$x = 0, y = 2.0$$
$$x = 1, y = 2.5$$
$$x = 2, y = 3.0$$
$$x = 3, y = 3.5$$

These pairs of x, y values are the *rectangular coordinates* of the points shown on the graph in fig. 1, through all of which the line passes. The points are established with reference to the intersection of the vertical axis (y) and the horizontal axis (x), which is called the *origin* (both variables having the value 0). Values of y above the origin are positive, while those below are negative. Similarly x values to the right of the origin are positive, those to the left are negative. Thus for the point with coordinates (3, 3·5) we move 3 units along x to the right of the origin and erect a perpendicular; 3·5 units above the origin establishes the position of y from which a horizontal is extended. The intersection of the vertical and horizontal lines fixes the location of the point on the graph. The coordinate of x will sometimes be called the *abscissa*, while y will be referred to as the *ordinate* of the point.

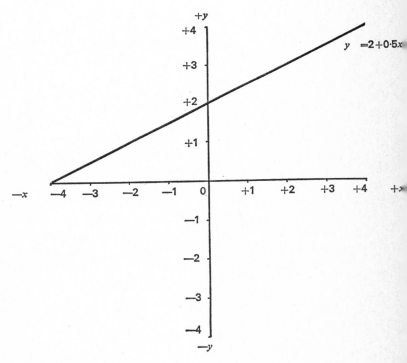

Fig. 1   Straight line.

Although the equation of the straight line is very common, we should remember that there are a multitude of equations which generate non-linear curves when graphed. The *quadratic* equation expressed by $y = a+bx+cx^2$ is one of these. If the constants a, b and c were 2, 1 and 0·3 respectively we would find that:

for $y = 2+x+0\cdot3x^2$, when $x = -3, y = 1\cdot7$
$$x = -2, y = 1\cdot2$$
$$x = -1, y = 1\cdot3$$
$$x = 0, y = 2\cdot0$$
$$x = 1, y = 3\cdot3$$
$$x = 2, y = 5\cdot2$$
$$x = 3, y = 7\cdot7$$

This situation is shown by fig. 2.

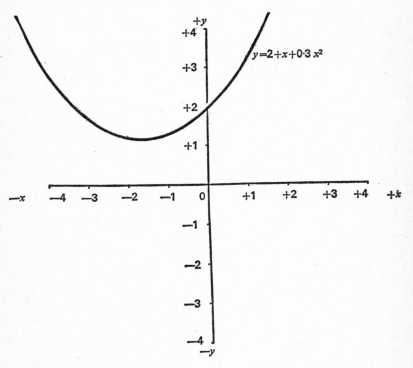

Fig. 2   Second-degree parabola.

Other shapes of quadratic equations emerge when different numerical values of the constants appear and when the signs of the constants are altered. In all cases of a *second-degree* equation (i.e. involving x^2) there will only be one bend in the curve. In the case of an equation of the third degree there will be two bends, of the fourth degree three bends and so on (see figs. 3 and 4).

21   The mathematical basis of statistics

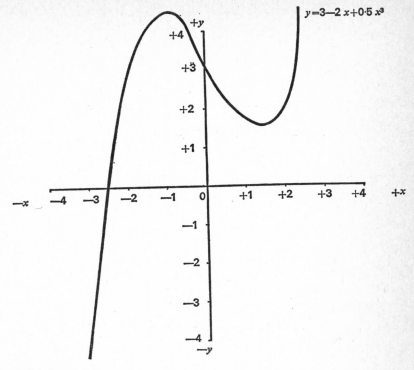
$y = 3 - 2x + 0.5 x^3$

Fig. 3 Third-degree curve.

Many equations which are fundamentally non-linear can be transformed into a linear form by various devices. More will be said about this later in the book, as these transformations can be of great value in the social sciences.

1.3.2 *Simultaneous equations*

1.3.2.1 One aspect of the statistical method makes considerable use of simultaneous equations. As a simple illustration of the techniques, suppose we have the following facts in front of us:

    (i) 8 gallons of petrol and 2 pints of oil costs 220 pence
    (ii) 4 gallons of petrol and 3 pints of oil costs 130 pence

From this data, how much does petrol cost per gallon and oil per pint? Letting *a* represent the cost in pence of a gallon of petrol and *b* the cost in pence of a pint of oil, we get:

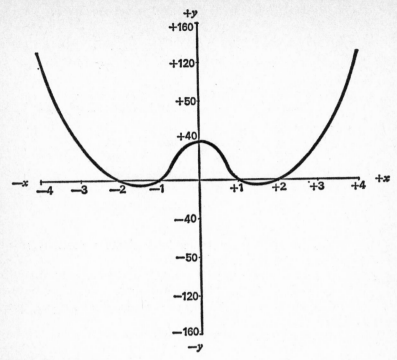

Fig. 4　Fourth-degree curve.

$$8a+2b = 220 \quad \ldots\ldots\ldots\ldots\ldots\ldots\ldots\ldots \quad \text{(i)}$$
$$4a+3b = 130 \quad \ldots\ldots\ldots\ldots\ldots\ldots\ldots\ldots \quad \text{(ii)}$$

The values of a and b must be such that these two equations are satisfied simultaneously. One method of solution is to find b in terms of a in equation (i) and substitute this in (ii):

thus
$$b = \frac{220-8a}{2}$$

giving
$$4a+3\,\frac{(220-8a)}{2} = 130$$

therefore
$$\frac{660-16a}{2} = 130$$

and
$$a = \frac{(130\times2)-660}{-16} = 25$$

Now substituting this value of a in (i) we can find b:

$$(8 \times 25) + 2b = 220$$

therefore $\qquad b = \dfrac{220 - 200}{2} = 10$

Petrol is 25p per gallon and oil 10p per pint.

The second method is to eliminate b by multiplying (i) by 3 and (ii) by 2 and subtracting:

$$\begin{array}{ll}
\text{(i)} \times 3 \text{ gives} & 24a + 6b = 660 \\
\text{(ii)} \times 2 \text{ gives} & 8a + 6b = 260 \\
\hline
& 16a + 0 \;\; = 400
\end{array}$$

therefore $\qquad a = \dfrac{400}{16} = 25$

This can again be substituted in (i) to find b. Basically, it is exactly the same method as before.

1.3.2.2 Finally, let us remember the one rule which must always be followed when manipulating an equation: the rule which stipulates that whatever is done to one side must be done to the other. For example:

$$\text{if } y = ax, \text{ then } \frac{y}{a} = x \text{ (division by } a\text{)};$$

$$\text{if } y = a + bx, \text{ then } y - a = bx \text{ (subtraction of } a\text{)}.$$

The same will apply where we take square roots or raise the two sides of an equation to a certain power.

1.4 Radicals, exponents and logarithms

1.4.1 We will have numerous occasions to deal with expressions involving square roots in statistics. The expression $9 = \sqrt{81}$ is a radical. Here 9 is the second (square) root of 81. $\sqrt{}$ is the radical sign and 81 is called the *radicand*. The effect of a radical symbol is exactly the same as that of a bracket. The complete expression has to be treated as a single number and therefore must be simplified before finding the root. Multiplication or division involving radicals is straightforward. In multiplication the radical of the product of the radicands provides the correct answer ($\sqrt{4} \times \sqrt{9} = \sqrt{\{4 \times 9\}}$) and in division the radical of the quotient of the radicands achieves a similar result $\left(\sqrt{36} \div \sqrt{4} = \sqrt{\dfrac{36}{4}} \right)$. Where non-radicals are involved in multiplication or division, multiply or divide the radicands by the square of the number $(8.\sqrt{25} = \sqrt{(64\times25)}$; $\frac{\sqrt{9}}{3} = \sqrt{\frac{9}{9}})$. Notice that before addition or subtraction can be undertaken the radical must be resolved:

$$\sqrt{4}+\sqrt{9} = 2+3 \neq \sqrt{(4+9)}.$$

Finally, remember that there is no real square root of a negative value.

1.4.2   As a convenient form of shorthand, we often write $10\times10\times10\times10$ as 10^4. This tells us that four tens are being multiplied together. More specifically, we should say that 10^4 is the 4th power of 10. The figure 4 is called the *exponent* of the power and 10 the *base*. There are certain definite rules about exponents which should be remembered. Any number raised to the power zero equals one. Thus $10^0 = 1$. From this, we can extend our shorthand by stating that $10^{-5} = \frac{1}{10^5}$, $10^{\frac{1}{2}} = \sqrt{10}$, and $10^{-\frac{1}{2}} = \frac{1}{\sqrt{10}}$. Multiplication and division follows the rules established for radicals: $8^2\times4^2 = (8\times4)^2$, $\frac{6^2}{8^2} = \left(\frac{6}{8}\right)^2$. Where the base is the same but the exponents differ, multiplication or division is achieved by addition or subtraction of the exponents: from this it follows that $14^{10}\times14^3 = 14^{10+3} = 14^{13}$, and $\frac{9^6}{9^4} = 9^{6-4} = 9^2$. Finally notice that $(4^2)^8 = 4^{2\times8} = 4^{16}$.

1.4.3   We know that $10^{-2} = \frac{1}{100} = 0.01$, $10^{-1} = \frac{1}{10} = 0.1$, $10^0 = 1$, $10^1 = 10$, $10^2 = 100$, $10^3 = 1000$ and so on. A unit increase in the exponent produces a tenfold increase in the value. But is it necessary to increase the exponent by a whole number in each case? Surely $10^{2.6990}$ will indicate some number between 100 and 1000 and $10^{-1.8350}$ some number between 0.1 and 0.01? This is in fact correct: a set of *common logarithms* shows the power to which 10 must be raised to produce a given answer.

The exponent has two parts: the integer or *characteristic* and the decimal fraction or *mantissa*. Tables of logarithms give only the mantissas, because the decimal part of the exponent to the base 10 of 0.5, 5, 50, 500, 5000, etc. will be identical. Only the characteristic will differ. As the mantissa is always positive (inferring the addition of some amount), it is related to the characteristic below. Thus 0.5 lies between 10^{-1} and 10^0, so that the characteristic is $\bar{1}$ (pronounced bar-one rather than minus one); 5 lies between 10^0 and 10^1 so that the characteristic is 0, and so on. As a mechanical rule, the positive characteristic will always be one less than the number of digits preceding the decimal point, while the negative (bar) characteristic will always be one greater than the number of zeroes immediately following the decimal place. This gives the following results:

Number		Characteristic
0·00783	$= 7·83 \times 10^{-3}$	$\bar{3}$
0·1800	$= 1·8 \times 10^{-1}$	$\bar{1}$
5·654	$= 5·654 \times 10^{0}$	0
10015·0	$= 1·0015 \times 10^{4}$	4

The establishment of the mantissa is obviously the main part of finding the logarithm of a number. The tables on pages 248 and 249 show the conventional layout of the mantissas. To find the log of any four-digit number involves three stages. The first two digits are identified from the extreme left-hand column. The appropriate mantissa in this row is that in the column headed by the value of the third digit. The mean difference corresponding to the fourth digit (found from the right-hand side of the page) is then added on to this mantissa. As an example, the mantissa of the logarithm of 2896 is 0·4609+0·0009 (the mean difference for 6) = 0·4618, with a characteristic of 3. For 0·00868 look up the mantissa for 868. Then the characteristic is chosen to take account of the two zeroes after the decimal point. If the number in question contains more than four digits, you must approximate. For 868295, look up the mantissa of 8683, but remember to take account of all the digits when fixing the value of the characteristic.

From the rules for dealing with exponents, it will be evident that to multiply two or more numbers together involves the addition of their logarithms, and the conversion of the answer back to an absolute form. Conversely, division involves only the subtraction of the logarithm of the divisor from the logarithm of the dividend. Squaring a number means doubling the logarithm of the number, while finding the square root involves halving the logarithm of the number. The conversion from the logarithm of the answer to the absolute answer is expedited by the use of antilogarithmic tables (pages 250 and 251), which are used in exactly the same way as log tables, and the reversal of the earlier principles to find the position of the decimal place. The procedure involved in addition, subtraction, doubling and halving the logarithms may, however, produce difficulties in determining the characteristic.

Where the logarithms have positive characteristics, there is no problem; proceed as in normal arithmetic. If, however, one or both of the characteristics are negative, some complication may be encountered. Suppose 85·36 is to be multiplied by 0·061. The logarithms are:

logarithm 85·36 = 1·9312
logarithm 0·061 = $\bar{2}$·7853
antilogarithm $\overline{0·7165}$ = 5·206

In this case, the 1 carried over to the characteristic from the mantissa is positive, and therefore added to the positive characteristic. Thus we have $2+(-2) = 0$. If 85·36 had been divided by 0·061, the subtraction would have been:

logarithm 85·36 = 1·9312
logarithm 0·061 = $\bar{2}$·7853
antilogarithm $\overline{3·1459}$ = 1399

Here the rule is to change the sign on the bottom line and add. The reason is the same as that for the subtraction of any minus quantity, i.e. $1-(-2) = 1+2 = 3$. Where a unit is carried over, this is a negative amount and so is added to the top characteristic, after which the above rule is followed. Suppose we have $\dfrac{0·2543}{0·09027}$:

logarithm 0·2543 = $\bar{1}$·4053
logarithm 0·09027 = $\bar{2}$·9555
antilogarithm $\overline{0·4498}$ = 2·817

Another problem is that of finding the square root of a number, the logarithm of which has a bar characteristic and is odd, i.e. logarithm 0·00352 = $\bar{3}$·5465:

$$2\overline{)\bar{3}·5465}$$
antilogarithm $\bar{2}$·7733 = 0·05933

The rule is to increase the characteristic by one bar unit so that it divides exactly by 2 (i.e. above $\bar{3}$ is changed to $\bar{4}$). This action is compensated by the addition of 1·0 to the first figure of the mantissa (0·5 becomes 1·5 which divided by 2 equals 0·7).

1.5   The summation sign

1.5.1   In statistics we often need to add (to sum) a series of figures. There is a shorthand form which tells us to do this. The symbol is the Greek capital letter sigma, written Σ. Suppose we have a series of figures designated x_i:

$$x_1 = 8$$
$$x_2 = 7$$
$$x_3 = 2$$
$$x_4 = 1$$
$$x_5 = 9$$
$$x_6 = 6$$
$$x_7 = 2$$
$$\overline{35} \quad \text{Then} \quad \sum_{i=1}^{7} x_i = 35.$$

The figures above and below the sigma sign tells us to sum from $i = 1$ (the first value of x) to $i = 7$ (the last value of x). In statistics the summation will very often be over all the values of the series in question, and in such cases it is generally unnecessary to include these limits of summation.

1.5.2   It should be noticed that in the example cited above x is a variable. Where we are concerned with a constant, a, then $\Sigma a = na$. For instance, the sum of $3+3+3+3+3 = 5 \times 3 = 15$.

What happens when the sigma sign precedes more than one term, i.e. $\Sigma(x+y-a)$ or $\Sigma(x \times y \times a)$?

In the first case, this can be expanded to $\Sigma x + \Sigma y - na$, showing that the summation of the algebraic summation of several terms is identical to the algebraic sum of the sums of the individual terms. The same approach cannot be extended to the second expression; $\Sigma(x \times y \times a)$ is not the same as $\Sigma x \times \Sigma y \times na$. The only acceptable modification is to take the constant to the outside of the summation sign, i.e.:

$$\Sigma(x \times y \times a) = a\Sigma(x \times y) \text{ or } a\Sigma xy$$

1.5.3   The problem is further complicated when the terms following the summation sign are squared:

$$\Sigma(x+y)^2$$

The bracket must be resolved before the principles mentioned in 1.5.2 can be used, so the solution cannot be $\Sigma x^2 + \Sigma y^2$. Instead:

$$\Sigma(x+y)^2 = \Sigma(x^2+y^2+2xy) = \Sigma x^2 + \Sigma y^2 + 2\Sigma xy$$

1.6   Significant figures, approximation and error

1.6.1   Numbers can be used in different ways: we count proper numbers but estimate common numbers. We can count the number of children in each family, or the number of houses subject to rates, but we can only estimate the height of a man or the speed of a car. Admittedly, the accuracy of the estimates in these two cases is increased by the employment of high-precision measuring instruments, but they are nevertheless estimates. Thus a man's weight is 11 stone (to the nearest stone) which means between 10 stone 7 pounds and 11 stone 7 pounds, or 11 stone (to the nearest pound) which indicates 10 stone 13 pounds 8 ounces to 11 stone 0 pounds 8 ounces. It seems that the concept of accuracy has two entirely different meanings. In dealing with proper numbers, a given result has no error; in dealing with measurable items, the accurate result is one lying within specified and required limits.

1.6.2   The problem of measurements should not be confused with that of approximation. Even when a countable result is available, we often prefer to round it off. Instead of saying that the output of cars from the motor industry is 1,973,816 vehicles, it is more comprehensible to give the figure as approximately two million. Strictly, one should define the degree of approximation. Thus we could express the above result as 2,000,000±50,000 or two million to the second significant figure, which means the same thing. The student will also have occasion to use the word 'significant' in dealing with decimal places. Thus 0·008632991 is 0·009 to one significant figure, 0·0086 to two significant figures, and so on (the significant figures start after the zeroes). The number of significant figures to be used will depend entirely upon the subsequent application of the answer. Expressed as a decimal, $\frac{1}{19}$ is 0·05263157 (to 7 significant figures). If we want to find $\frac{1}{19}$ of certain amounts of money, accurate to the nearest penny, then the number of significant figures varies with the size of the amounts of money involved. Thus:

0·0053	of £1·52½	= 8p to nearest halfpenny.
0·00526	of £28·75	= £1·51 to nearest halfpenny.
0·005263	of £100	= £5·26½ to nearest halfpenny.
0·0052632	of £1,000	= £52·63 to nearest halfpenny.

The principal pitfall in using approximate figures is that of giving the impression of greater precision than actually exists. If we add up the following:

215	(accurate)
2,500	(to the second significant figure)
675,000	(to the third significant figure)
677,715	

the answer we get is most misleading. In reality, the second value could be anywhere between 2,450 and 2,550, and the third anywhere between 674,500 and 675,500. Therefore, instead of 677,715 the answer might be between

215	215
2,450	2,550
674,500	675,500
$\overline{677,165}$ and	$\overline{678,265}$

This is a difference of ± 550. It would have been better to state the original answer in this form. Similarly, to suggest that 0·042 (to the second significant figure) divided by 0·056232 (to the fifth significant figure) is equal to 0·74691 is incorrect.

1.6.3   When we use some form of approximation, such as $a \pm x$ or a to the nth significant figure, we are admitting that the answer may be in error. At first sight it may seem to be perfectly adequate to indicate the error in absolute terms, but a simple example will show that a relative expression of the error in question is preferable. Suppose that two men estimate their expected annual salary in five years' time; the first suggests that it will be £1,600 and the second £4,500. After the five years have elapsed, the actual figures are £1,500 and £4,600. The absolute error in each case has been £100, yet the second estimate was better than the first. The relative errors were:

$$\frac{100}{1500} = 0\cdot067 \text{ or } 6\cdot7\%, \text{ and } \frac{100}{4600} = 0\cdot022 \text{ or } 2\cdot2\%$$

1.7   Examples

1.7.1   Simplify the following expressions:

(a) $8 \times 6 + 5 \div 3 - 6 + 5$
(b) $(4 \times 12) \times (6 + 2) + 8 \times 3$
(c) $[(5 + 6)(3 \times 2) + (7 - 6)] \times (7 \div 3)$
(d) $-8 + (6 - 4 + 2 - 1) + 18 - 4 - 2$
(e) $(-3) + (-6) + (-14) + (18) + (2) + (-3)$
(f) $(0)(0 + 18) - (0 \div 2)$
(g) $(14 + 2 \times 6 - 3)(1 - 1 \times 3) \div (6 - 8 \times 3 + 2)$
(h) $-15 - (-3)$
(i) $(-3 \times 6) \div (-4 \times -6 \times -3 \times -1)$
(j) $8[(54 - 3)(6 + 2 - 1)] \times -5$

1.7.2 Simplify and evaluate the following:

(a) $\frac{1}{8}+\frac{1}{4}-\frac{6}{3}+\frac{14}{5}$

(b) $2\frac{3}{8}-1\frac{6}{7}$

(c) $\frac{14}{1}\times\frac{2}{7}\times\frac{9}{5}\times\frac{5}{10}$

(d) $\frac{19}{20}\div\frac{6}{8}$

(e) $\left(\frac{7}{9}+\frac{4}{3}\right)\div\left(\frac{2}{5}-\frac{1}{12}\right)$

(f) $1\cdot835\times0\cdot00031$

(g) $0\cdot012+8\cdot0+371\cdot85+0\cdot15+0\cdot00089$

(h) $0\cdot875\div16\cdot3825$

(i) $(0\cdot5\times-\cdot75)\div(8\cdot35\times0\cdot003)$

(j) $\frac{1}{5}\times0\cdot067-\dfrac{3\times0\cdot25}{\frac{1}{4}}$

1.7.3 Draw the graphs of the following:

(a) $y = 2\cdot4-1\cdot75x$

(b) $3\cdot6 = 5y-2x$

(c) $y = 3\cdot0+0\cdot8x+x^2$

(d) $y = \dfrac{1}{x^2}$

(e) $y = x^2-x^3$

Solve the following simultaneously:

(f) $15\cdot3 = 18a+b$
  $6\cdot0 = 1\cdot5a+3b$

(g) $a = 2b+6$
  $-0\cdot5 = 7a+3b$

(h) $15 = 10b$
  $a = 14c+15-b$
  $12 = a+c$

1.7.4 Simplify and evaluate:

(a) $(\sqrt{3}\div\sqrt{6})\times5^2$

(b) $(8\times9^2)\div\sqrt{25}$

31 The mathematical basis of statistics

(c) $8^2 \times 8^5 \div 8^3$

(d) $\sqrt{6} \times \sqrt{6} \times 36$

(e) $\dfrac{\sqrt{(14+16+3)}}{\sqrt{9}+\sqrt{196}+\sqrt{256}}$

Calculate the following, using logarithms:

(f) $18 \cdot 63 \times 15 \cdot 82 \times 0 \cdot 03$

(g) $0 \cdot 0053 \div 6 \cdot 25$

(h) $(9652 \cdot 0 \div 28 \cdot 23) \times 6 \cdot 321$

(i) $15 \cdot 86^2 \div \sqrt{0 \cdot 00816}$

(j) $1754^2 \div \sqrt{0 \cdot 00717}$

1.7.5   From the following:

x	y	a
8	15	7
3	7	7
12	3	7
7	12	7
2	9	7
1	6	7

calculate:

(a) $\Sigma(x+y)$

(b) $\Sigma(x+y)^2$

(c) Σxy

(d) Σxy^2

(e) $\Sigma(xy)^2$

(f) $(\Sigma xy)^2$

(g) $\Sigma(a+y)$

(h) $\Sigma a(x-y)$

(i) $\Sigma xy - \Sigma x \Sigma y$

(j) $na - (\Sigma x \Sigma y)^2$

1.7.6   What do the following equal?

(a) $(145 \pm 6) + (92 \pm 19)$

(b) $(58 \pm 3) - (51 \pm 5)$

(c) $(18 \pm 7) \times (17 \pm 1)$

(d) $(33 \pm 5\%) \div (19 \pm 10\%)$

(e) $(82 \pm 5) \div (16 \pm 10)$

Chapter 2
The arrangement and presentation of data

Applied statistics is concerned with numerical data of various sorts. The economist may want information about earnings, the sociologist is interested in the structure of families, the educationalist requires facts about aptitudes, the businessman's needs involve a knowledge of profit margins. These requirements can only be met by research.

Often no data are available and a survey may be needed, in which people or institutions are asked to provide information about themselves. The questions may be put by interviewers, or through the medium of postal questionnaires. Sometimes the respondent cannot provide an answer and the sponsor of the survey must accept responsibility for the necessary measurement himself; this could involve either simple determination of heights, weights, areas, etc., or complex tests to evaluate achievement or potential. Even if data are available, they may exist in an unsuitable form, or may have to be extracted by sorting through past records, sifting masses of tabulated figures, and amalgamating and condensing data originally collected in another context.

Whether the raw material is primary data (that collected at source) or secondary data (which exists already in reports, balance sheets, government publications and so on) it is certain that it will need arranging and probably presenting in some way before it can be appreciated and interpreted. A pile of completed questionnaires, or a long list of regional sales figures, have no immediate impact; they convey nothing to the statistician or to anyone else. The answers on the questionnaires must be taken separately, the different responses being enumerated; the sales figures must be broken down to provide information about the performance of different products and different salesmen at different times.

From the apparent chaos must appear an ordered pattern developed according to logical principles.

2.1 Tabulation

2.1.1 The most convenient method of producing an ordered pattern of numerical data is the construction of a table. Tables can be built up in many different ways, but they all possess the same fundamental attribute: it is possible to incorporate in them a pattern over space, over time or with respect to size. We will examine several basic types of table, but before doing this let us look at certain rules about the construction of tables in general.

2.1.2 Tabulation is used to facilitate the understanding of complex numerical data. A table should therefore be as simple and unambiguous as possible, the title and column or row headings being brief but self-explanatory. The units of measurement employed must be clearly shown and sufficient space should be allowed to range units under units, tens under tens etc. Approximations and omissions can be explained in footnotes, together with information about miscellaneous items and changes in methods of measurement over time.

In addition to these points relating to definition, the general format of the table must be considered. It should not be so large that it confuses the reader, for this defeats the object of the tabulation. A vertical rather than a horizontal arrangement is preferable, although not invariably possible, as the eye moves down columns more readily than along rows. Sets of data which are to be compared should be close together, and percentages and averages ought to be beside the figures to which they relate. Distinct sections of a table can be distinguished by the use of heavy lines or double lines, while important figures can be made to stand out by underlining, using different colours or employing a bolder type.

2.1.3 As an illustration of these principles, consider a firm running a group of supermarkets that wishes to bring together all its sales data into one table for incorporation in a shareholders' sales report. The table might well look like fig. 5. It is based upon the assumption that a progress report over the year in question is important. If we had wanted to stress the relative importance of the three departments, then a vertical rather than horizontal presentation of these sub-sections might be used.

Additional features which could be incorporated include:

(i) Percentage figures of departmental sales to total sales for each month. These could be placed in brackets after the absolute figure.

Receipts from sales (£ thousands) 1 April 1965 to 31 March 1966

	All Depts	Grocery Dept						Butchery Dept				General Household Goods				
	(1)	(2)	(3)	(4)	(5)	(6)		(7)	(8)	(9)	(10)	(11)	(12)	(13)	(14)	(15)
Month	Total	Dept total	Green-grocery	Dairy produce	Frozen and tinned foods	Confec-tionery		Dept total	Fresh meat	Bacon, ham etc.	Cooked meat	Dept total	Soap, powder etc.	Station-ery	Garden-ing	Miscel-laneous
April																
May																
June																
July																
August																
September																
October																
November																
December																
January																
February																
March																
Total																
% (totals)		(2) of (1)	(3) of (2)	(4) of (2)	(5) of (2)	(6) of (2)		(7) of (1)	(8) of (7)	(9) of (7)	(10) of (7)	(11) of (1)	(12) of (11)	(13) of (11)	(14) of (11)	(15) of (11)

Fig. 5 XYZ supermarket chain.

(ii) Each column and row might be numbered for reference purposes in the report. Identification of an individual figure would be greatly facilitated in this way.

(iii) Departmental and grand totals might be produced in larger print so as to make them stand out.

The precise form of presentation will, of course, depend entirely upon the requirements of the user(s). In any tabulation of this general type the person designing the table should be in continual contact with other interested parties, so that all aspects of the use and purpose of the table can be explored.

2.2 The frequency distribution

2.2.1 In a survey of fifty families conducted to establish the structure of the family unit in a certain area, one of the questions asked was 'how many occupants live in this household?' The following list, extracted from the completed questionnaires, provides the answer:

Number of occupants in fifty households

4	7	4	6	4
2	3	6	3	5
6	3	17	9	8
1	3	4	2	2
1	11	3	8	5
6	4	5	14	4
3	2	1	4	6
5	6	7	15	3
4	4	4	8	4
2	3	5	4	4

This is a form of tabulation, but a most unsatisfactory one. The eye fails to see any structure in this situation. The number of occupants obviously varies from house to house, but whether one figure predominates is not immediately obvious. What is apparent is that most of these figures occur more than once. If we examine the results, find the highest and lowest figures, and finally record how many times each intermediate value appears, we are constructing a *frequency distribution*.

The very term, frequency distribution, explains itself. There are obviously a number of different occupancy values; there is a 'distribution' of figures, and each one occurs a certain number of times (i.e. with a given frequency).

Number of occupants	Number of houses
1	3
2	5
3	8
4	13
5	5
6	6
7	2
8	3
9	1
10	0
11	1
12	0
13	0
14	1
15	1
16	0
17	1
	50

Although this gives a better impression of the survey's results, showing that the most common number of house occupants is 4 with the majority of the rest clustering closely round this figure, it is still rather cumbersome. There is a considerable amount of detail for the eye to take in and a condensed tabulation would be infinitely better. This can be achieved by grouping some of the figures together into class intervals, and noting the frequency of the combined values.

Number of occupants	Number of houses
1 or 2	8
3 or 4	21
5 or 6	11
7 to 10	6
11 to 15	3
16 and over	1
	50

Admittedly, some of the definition has been lost, but there is a compensation. An immediate awareness of the structure of occupancy is now possible. There is a rapid build-up to the most common group of 3 or 4 occupants (probably parents plus small families) with a steady decline to the larger numbers of occupants (possibly large families, including several children, parents, grandparents etc.).

2.2.2   The reader may ask how it is possible to decide what the groups should be and how far the grouping can be taken. The decision largely rests upon experience. The groups may be relatively large if there are few occurrences of value, but greater detail in the form of smaller class intervals is needed over those ranges having high frequencies. The actual extent to which the condensing of the data is undertaken generally rests upon common sense, and just as the first table had too much detail, so it is possible to progress to the other extreme:

Number of occupants	Number of houses
1 to 6	40
7 to 15	9
16 and over	1
	50

In all circumstances, there would appear to be a happy medium.

2.3   Variables

2.3.1   We noticed that the number of occupants in the fifty houses varies quite considerably. Therefore we say that occupancy is a variable. We may define a variable as a feature characteristic of any member of a group, yet differing in quantity from member to member. There are fundamentally two types of variable. Some variables can only take integer values: there is a definite jump from one value to another. The number of occupants in a house can be 1 or 2 or 3 etc.; it cannot be $1 \cdot 32$ or $3 \cdot 86$. Variables of this kind are described as *discrete*. Many other examples can be seen to exist, such as the number of shareholders in a group of limited companies, the earnings of capstan lathe operators (the discrete steps are pence in this case) and the number of children in each class in a school; all illustrate the features of the discrete variable.

The reader may see a resemblance here to the comments made in 1.6.1, where we distinguished between proper and common numbers.

The discrete situation involves the use of proper numbers. How then are we to describe the variable which rests upon common numbers? Previously we stated that we could only estimate the height of a man or the speed of a car (say to the nearest $\frac{1}{2}$ inch or 1 m.p.h., respectively), because one man can differ in stature from another by an infinitely small amount and the speed of a car may vary by a tiny fraction of 1 m.p.h. Both are variables which can exhibit any value between two limits; there is a continuous scale of measurement, so we describe the variable as being *continuous*.

In practice, continuous variables must be converted to a discrete form by expressing values to the nearest appropriate unit of measurement. No measuring instrument can possibly distinguish infinitesimally small differences. By definition they exist, but cannot be measured, except by means of the differential calculus.

2.3.2   Many different methods are employed to show the variable range of the class intervals in a frequency distribution. Let us briefly examine the implications of some of these. Where a discrete variable is involved, there should be no problem at all.

Number of cartons packed by operative during one shift (x)
100–199
200–299
300–399
400–499

This means exactly what it says. Unfortunately, the reader may encounter the following forms:

x		x
100–200	or	100–
200–300		200–
300–400		300–
400–500		400–

Here one must realize that a discrete variable is shown and infer the first presentation.

The difficulty is in another form when continuous variables are involved. The following presentations may seem to be perfectly valid:

Weight of tins of coffee

8·0 oz. but less than 8·2 oz.
8·2 oz. but less than 8·4 oz.
8·4 oz. but less than 8·6 oz.

or
8·0 oz. to 8·2 oz.
8·2 oz. to 8·4 oz.
8·4 oz. to 8·6 oz.

or simply
8·0 oz.–
8·2 oz.–
8·4 oz.–

They certainly all show continuous variables, but what they do not show is the extent of the estimation used. Are the measurements exact to the nearest hundredth or nearest tenth of an ounce? We just do not know. We must infer that 8·0 oz. but less than 8·2 oz. means 8·0 oz. up to and including 8·19999' oz. In other words, we treat the measurements as being exact, even though they could not have been so in practice. It is preferable to state the degree of approximation explicitly or to use the following version:

Weight of tins of coffee

8·0 oz. to 8·1 oz.
8·2 oz. to 8·3 oz.
8·4 oz. to 8·5 oz.

In this case, we are immediately made aware of the fact that measurement has been to the nearest tenth of an ounce.

2.3.3   Another feature of class interval construction which often worries students of applied statistics is the presence of open-ended class intervals. For example:

Age of holders (years) of driving licences

Under 20
20–29
30–39

Number of shares in company
1–   499
500–   999
1,000– 4,999
5,000– 9,999
10,000–99,999
100,000 and over

In the first example, there is a legal restriction on the possible lower boundary of the first class interval. If the distribution is confined to car drivers, then it is 17 years; if motor-cyclists are included, then it is 16. Very often the law, physiology, or common practice prescribes the answer in this way, but it is not invariably the case as the second illustration shows. Without some prior knowledge of the shareholding structure of this particular company it is impossible to be certain about the upper boundary of the final class interval. The sky would seem to be the limit.

All that can be done here is to examine the pattern of class intervals throughout the distribution. If they have been increasing as in this case, then the increase should be maintained. The actual magnitude of the increase will depend on the number of shareholders in this group. If it is small then it would be risky to infer a very large increase. If, on the other hand, it is large, the upper limit may be set at a higher level. The reader should be aware that these are only general principles, and cannot replace a detailed knowledge of the situation.

2.3.4 *Relative frequency and cumulative frequency*

2.3.4.1 We have been concerned so far with the problem of grouping and the presentation of the class intervals which show the groups. Let us now turn our attention to alternative methods of arranging the frequency figures.

One difficulty often encountered is the comparison of results from two investigations, either undertaken in different places or at different times, where the total number of observations is not the same. Consider the two distributions below, showing the number of calls per day received by the servicing department of a central-heating firm.

The earlier investigation took results from 24 working days, while that for 1965 was over a whole year. Whether there has been any significant change is not at all apparent because of the differences in the absolute frequency values. What is needed is a means of reducing them to a common basis, and the simplest way of doing this is to express

Number of calls	1960 Number of days	%	1965 Number of days	%
0–10	6	25	60	19
11–20	12	50	168	54
21–30	4	17	64	20
31–40	1	4	12	4
41–60	1	4	8	3
	24	100	312	100

each frequency figure as a percentage of the total. The effect of doing this is shown above. We have produced a relative frequency table in each case. Direct comparison is now possible, yet the relationship between the different groups has been maintained: 6: 12: 4: 1: 1 is almost the same as 25: 50: 17: 4: 4.

2.3.4.2   Another aid to interpretation is found in the construction of a cumulative frequency table in which we ask how many (or what percentage) of the observations lie above or below a given value. Using the relative frequency distribution above gives:

Number of calls	Number of days 1960 % cumulative frequency	1965 % cumulative frequency
0–10	25	19
11–20	75	73
21–30	92	93
31–40	96	97
41–60	100	100

By adding each successive frequency figure to the previous cumulative figure, we find that 25% of working days have up to and including 10 calls for service, 75% have up to and including 20 calls per day, and so on. We could also look at this in another way:

| Number of calls | Number of days | |
	1960 % cumulative frequency	1965 % cumulative frequency
0–10	100	100
11–20	75	81
21–30	25	27
31–40	8	7
41–60	4	3

Here we see the percentage of days which have had more than so many calls. This version might therefore be described as a *more than* cumulative frequency table, as opposed to the first which is the *less than* cumulative frequency case.

2.3.5 Attributes and variables

Frequency distributions are not uniquely concerned with variables. It often happens that differences between one person and another, one house and another etc. are qualitative rather than quantitative. This sort of manifold classification can nevertheless be considered as a frequency distribution, as is illustrated below:

Socio-economic group	Number of families
Higher managerial, senior professional etc.	18
Other salary	42
Skilled manual	61
Semi-skilled manual	59
Unskilled manual	32
	212

The difference between one group and another in these circumstances is the possession or non-possession of a prescribed attribute rather than a difference in magnitude due to the variable nature of the situation under review. The analysis of results in this form is considered in detail in *Statistics for the Social Scientist:2 Applied Statistics*.

2.4 Histograms, polygons and frequency curves

2.4.1 Although a correctly constructed frequency distribution table is reasonably easy to comprehend, the eye can very often appreciate the implications of a situation more readily when a diagrammatic presentation is used. For instance:

Number of customers served in one day by shop assistants	Number of assistants
8	4
9	6
10	8
11	18
12	15
13	11
14	5

Why not in this case show the different frequency figures by lines of different lengths, as in fig. 6?

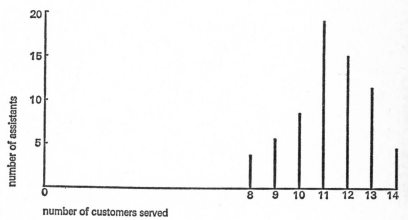

Fig. 6   Line chart.

This simple diagram, called a line chart, clearly indicates that the majority of assistants are serving eleven or more customers. This is seen from the fact that the longest lines are clustered to the right of the variable range.

2.4.2   An extension of the principle providing for the representation of variable values (x) along the horizontal axis of a 'graph' and the frequency (f) of each value's occurrence on the vertical axis can similarly be used to show a grouped frequency distribution.

Hire-purchase debt outstanding £	Number of families
Less than 100	28
100 but less than 200	32
200 „ „ „ 300	17
300 „ „ „ 400	9
400 „ „ „ 500	6
500 „ „ „ 600	1
	93

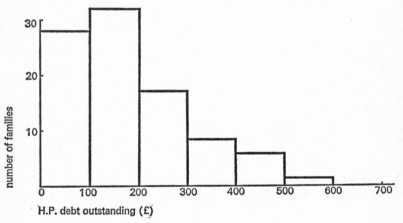

Fig. 7   The histogram.

In fig. 7 we are using a block rather than a line to represent the frequency of each class interval, and we produce what is termed a *histogram*. It may seem that this is the only difference, but there is one important feature of the histogram which should always be borne in mind. Whereas the length of the lines in fig. 6 are proportionate to the frequencies, it is the areas of the blocks which must be proportionate to the frequencies in the histogram. This can easily be seen if we amalgamate the last two groups in the distribution of outstanding hire-purchase debt, giving a frequency figure of 7. If the new block with a base double the numerical width of the others is given seven units of height, it will be completely out of proportion to the original histogram. For the areas of the blocks to maintain their proportionate relationship with the frequencies, the block height must be halved as the numerical width of the block is doubled (fig. 8).

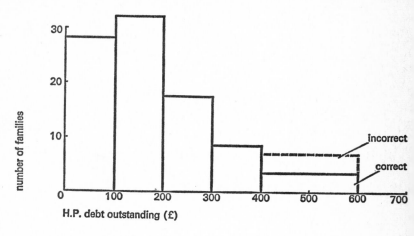

Fig. 8

2.4.3 *The frequency polygon and curve*

Instead of drawing the histogram of a distribution in full, it is often more convenient to produce a *frequency polygon*. This involves no more than joining up the midpoints of the upper extreme of the blocks with straight lines as is shown in fig. 9.

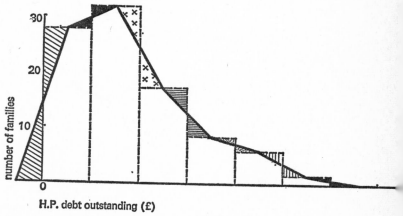

Fig. 9   Frequency polygon: distribution of hire-purchase debt.

The reader will notice that in the case of the first and last class intervals the line has been extended beyond the original range of the variable. The reason is that the area under the polygon should be the same as that in the histogram (i.e. the total frequency). Only if each triangle cut off the histogram is compensated for will this requirement be met. This can be seen from the correspondingly shaded triangles in fig. 9.

The student of the statistical method will sometimes encounter a *frequency curve* instead of a polygon.

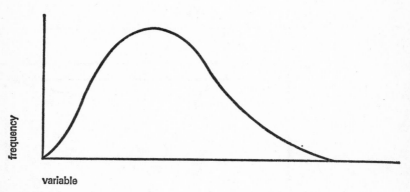

Fig. 10   Frequency curve.

This will never be produced from an actual situation of the real world. It is not difficult to see that a continuous variable is involved, but how can we possibly measure the absolute frequency of an infinitesimally small range of the variable? The answer is that we can't; we are looking at a distribution which has existence only in the theory of statistics. We must not imply, however, that a theoretical distribution of this sort is of no use, for very close approximations to theoretical distributions are often generated in the real world: so close in fact that it is quite valid to utilize their properties, as developed by mathematicians, to aid analysis in the problem being considered.

2.4.4   *The ogive curve*

In just the same way that the frequency distribution table can be presented graphically, there is a visual form of the cumulative frequency table. This is shown for the data below (fig. 11).

Test life of dry cell batteries (minutes)	Number of batteries tested	Cumulative frequency less than	Cumulative frequency more than
0		0	100
20	3	3	97
30	8	11	89
40	17	28	72
50	30	58	42
60	23	81	19
90	10	91	9
120	9	100	0
	100		

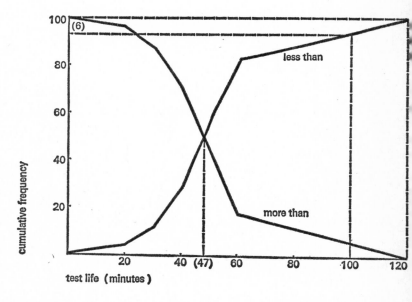

Fig. 11   Ogive curves.

A number of points should be noted in connection with these graphs. Firstly, each cumulative frequency figure should be plotted against the upper boundary of the class interval ('less than' cumulative) or the lower boundary ('more than' cumulative); the midpoints of the class interval should never be employed for this purpose. Our knowledge of the purpose of cumulative tables should tell us this. Secondly, the decision to use straight lines to join up the points rather than a smooth curve rests upon the assumption that the observations in each group are evenly scattered between the two boundaries of the class interval. This is an assumption which we shall usually adopt when calculation problems are encountered in the next chapter.

Finally, we should be aware that from the ogive curve we can estimate the number of observations which lie between given limits of the variable (other than those shown in the original classification). For instance, we can find that half the batteries tested have a life of less than 47 minutes while half have a life of more than 47 minutes. (This is the point where the two ogive curves intersect.) Similarly, we can ascertain the number of batteries tested which had a life between 100 and 120 minutes. We simply draw up vertical lines from these values; at the point of intersection with one of the ogives, horizontal lines are extended to the cumulative frequency axis. The numerical difference between the two readings on this axis provides the answer of 6 (see fig. 11).

2.4.5 *The Lorenz curve*

Let us end this section by looking at one particular application of the principles so far discussed. The modern world has seen the development of a trend towards greater equality in the distribution of wealth and particularly of income. This has been particularly noticeable in the United Kingdom, where fiscal policy has reduced inherited fortunes and gross earnings. How can we ascertain whether in fact the distribution of income has changed over time, or whether the incidence of taxation through its progressive form has brought about greater equality? One approach is to ask whether 10% of income earners are earning 10% of the total income, whether 50% of earners receive 50% of the total income and so on. If there is perfect equality in the distribution of income, then there will be this one-for-one relationship. If not, and say 50% of earners receive only 35% of the income, then there is an obvious indication of inequality. The construction of a *Lorenz curve* gives an immediate impression of the actual situation that confronts us.

The data considered below are taken from *National Income and Expenditure* 1963 and relate to 1962. The steps in the calculation are self-explanatory. Fig. 12 shows the graph plotted from the results, demonstrating that even after tax there is still considerable inequality in the United Kingdom. However, it is worth noticing that the 'area of inequality' would have been considerably more extensive if data for income 'before tax' had been used. In general, the Lorenz curve is used as a comparative tool in statistics, rather than for providing a quantitative measure of inequality.

Income after tax	Numbers (000's)	Midpoints	Income of groups (thousands)
Less than £250	1407·0	125	246,100
£250 but less than £500	6293·0	375	2,360,000
£500 but less than £750	6672·0	625	4,170,000
£750 but less than £1,000	4400·2	875	3,850,000
£1,000 but less than £2,000	2493·0	1,500	3,740,000
£2,000 but less than £4,000	297·0	3,000	891,000
£4,000 but less than £6,000	33·7	5,000	168,500
£6,000 and over	4·1	15,500	63,550
	21600·0		15,489,150

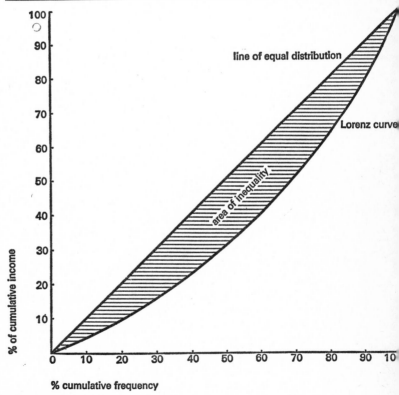

Fig. 12   Lorenz curve.

Cumulative frequency	Percentage cumulative frequency	Cumulative income of groups	Percentage cumulative income of group
1407·0	6·51	246,100	1·59
7700·0	35·65	2,606,100	16·83
14372·0	66·54	6,776,100	43·75
18772·2	86·91	10,626,100	68·60
21265·2	98·45	14,366,100	92·75
21562·2	99·82	15,257,100	98·50
21595·9	99·98	15,425,600	99·59
21600·0	100·00	15,489,150	100·00

2.5 Time series

In previous sections, our tabulations of variable data have largely been based upon quantitative patterns. Surely, however, temporal patterns can have relevance to us? The sociologist, the economist, and the businessman are all particularly interested in variable movements over time. A *time series* tabulation, therefore, is based upon chronological order; it studies a single variable quantity at successive points or intervals of time. By and large, the reader will be familiar with time series tables showing, for instance, increases or decreases in gold and convertible currency reserves, changes in the cost of living, or the profits of a company. These are frequently found in the various media of communication to the general public. In consequence, we shall omit a separate discussion of tabulation procedures here, but adequate illustrations will be provided when the various methods and difficulties of graphing time series data are considered.

2.5.1   Let us first look at the main types of time series data which we are likely to encounter. Suppose that one measures the mechanical efficiency of the shock absorbers on a car at six-monthly intervals from the time the car is new, so that the following results are obtained:

Age of car (years)	Efficiency of shock absorbers (per cent)
New	100
0·5	99
1·0	97
1·5	95
2·0	92
2·5	88
3·0	83
3·5	75
4·0	67
4·5	60

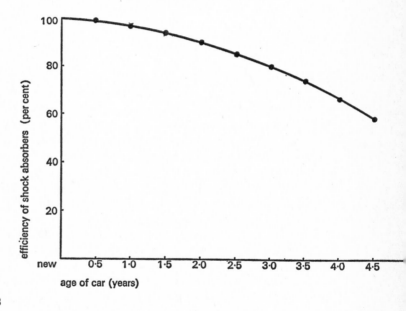

Fig. 13

In fig. 13 it would seem logical to use a curved line to join up the points, as the wear on the shock absorbers is increasing continuously. For instance, it is certain that their efficiency is not going to improve between any two test periods and it is most unlikely that there will be sharp increases in wear at particular points between the half-yearly inspections.

Exactly the opposite comment applies when one considers a variable such as bank rate:

Time		Bank rate (per cent)
July	1961	7·0
October	1961	6·5
November	1961	6·0
March	1962	5·5
March	1962	5·0
April	1962	4·5
December	1962	4·0
February	1964	5·0
November	1964	7·0

Here the changes are sudden, taking place at a specific moment in time. They should therefore be shown as in fig. 14.

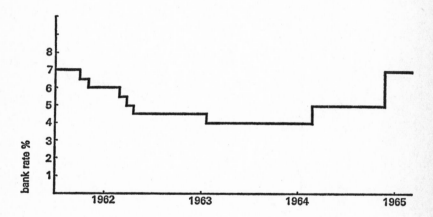

Fig. 14

The most common type of time series data is that involving monthly, quarterly or yearly aggregates, whether these are sales, output, population, savings or any one of many other possible variables. Consider gross national product, shown below:

Year	G.N.P. (at 1958 constant prices) (£ million)	G.N.P. percentage change compared with 1958
1958	21,768	0
1959	22,628	4·0
1960	23,719	9·0
1961	24,517	12·6
1962	24,762	13·8
1963	25,911	19·0
1964	27,289	25·4

(*National Income and Expenditure* 1965)

Plotting the percentage changes for convenience (fig. 15), we join up the points by straight lines.

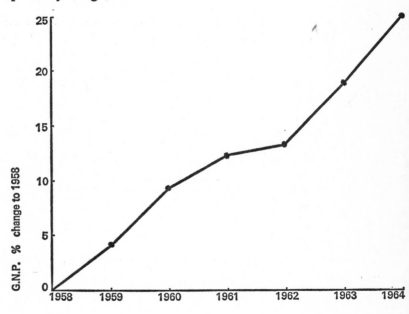

Fig. 15

The reason for this is self-evident. These are figures which neither show continuous change nor sporadic and immediate change. The points are joined up merely as an aid to visual interpretation, making the pattern of movements in the aggregates from year to year immediately apparent.

2.5.2   The graphing of the time series in 2.5.1 produced no real difficulty. Problems are, however, frequently encountered in practice, so we shall now look at some of these and suggest possible solutions.

The first group of problems concern the scale of measurement and the position of the axes:

Year	Employment as percentage of total (June)
1955	99·0
1956	99·0
1957	98·3
1958	98·1
1959	98·2
1960	98·6
1961	98·8
1962	98·3
1963	98·0

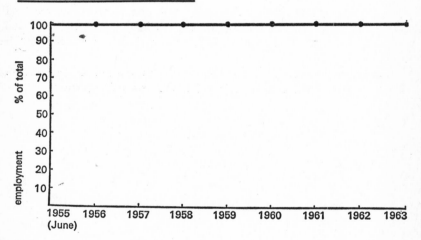

Fig. 16

Plotting this in a conventional manner (fig. 16) produces a graph of dubious value. On the one hand it is almost impossible to detect the changes in the level of employment, while on the other there is a considerable amount of wasted space. One solution would be to graph unemployment (i.e. 100 minus each employment percentage). Not all situations are capable of transformation in this way. What is required is a general method of solution, as shown in fig. 17.

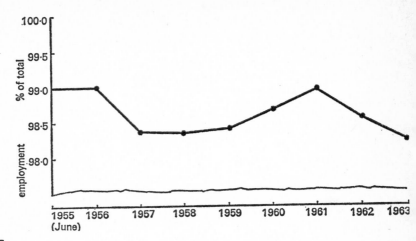

Fig. 17

In this instance the suppression of the origin has been indicated by a jagged edge at the bottom (representing a tear), but any suitable device may be employed to indicate the omission of a section of the vertical axis.

2.5.3   Sometimes it is necessary to show a break in the time scale rather than in the variable scale. For the data below this is true:

Year	U.K. consumers' expenditure abroad (£ million)
1938	31
1948	94
1949	105
1950	116
1951	152

(*National Income and Expenditure* 1963)

There could be no justification for joining up either the points for 1938 and 1948 or the horizontal axis in fig. 18. Not only is the time interval ten times greater than over the rest of the graph, but there must inevitably have been a drop in expenditure during the war years. The break should suggest these facts to the person studying this situation.

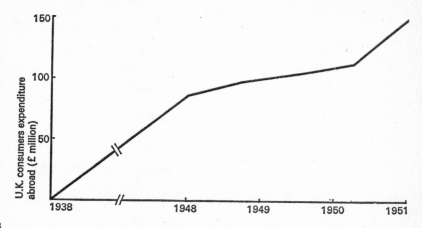

Fig. 18

There should also be a break in the graph when the method of compiling or calculating the figures under review has been changed. Such is the case when looking at the following series, which shows the index of retail prices:

Year	Month	Index of retail prices Jan. 1956 = 100	Jan. 1962 = 100
1961	September	115·5	
	December	117·1	
		. . . . .	
1962	March	118·1	100·5
	June	120·9	102·9
	September	122·4	104·1

(*Monthly Digest of Statistics*, January 1963)

Although it is quite possible to calculate figures for 1962 in terms of January 1956 = 100, the January 1962 = 100 series is nevertheless effectively current weighted while the earlier one was base weighted. This information should be included in a footnote (fig. 19).

57 The arrangement and presentation of data

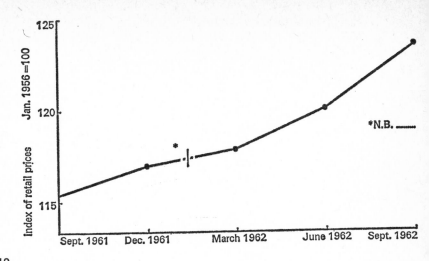

Fig. 19

2.5.4   While it is true that a straightforward visual impression of the changes in a single variable can be of the greatest value, it is equally true that the graphical comparison of two or more time series may be of considerable assistance in finding relationships between the several variables. Such comparisons are easily presented if the variables of the

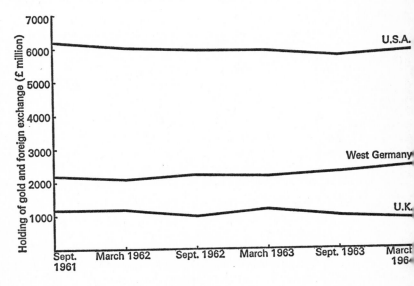

Fig. 20

different series are of the same nature and given in the same units (fig. 20).

Year	Month	Official holdings of gold and foreign exchange (£ million) U.K.	U.S.A.	Germany
1961	September	1,269	6,257	2,299
1962	March	1,233	6,026	2,175
	September	997	5,904	2,310
1963	March	1,005	5,742	2,302
	September	977	5,639	2,454
1964	March	950	5,711	2,541

(*Barclays Bank Review*, August 1964)

One word of warning should be given at this point. Even where the units of measurement are the same, the gradients of the lines joining up points may give a completely erroneous impression of the facts, as fig. 21 shows.

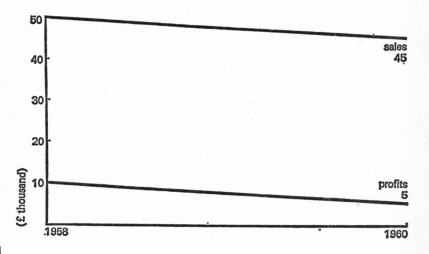

Fig. 21

In 1958 the profits represent 20% of sales. As the gradients of the profits and sales lines are identical, it may appear that the same applies in 1960. This is not the case. Profit as a percentage of sales is now 11·1. There has been a significant change in the profit margin of this hypothetical firm. In any natural scale presentation of data only absolute changes can be considered: a semi-logarithmic presentation would be required if we wanted to find out if the proportionate relationship between the variables had been maintained (this will be considered in the next section).

A complication arises in making comparisons when the units of measurement of the two series are different.

Year	Unfilled vacancies (monthly averages, 000's)	Percentage of unemployment (monthly averages)
1958	197·6	2·1
1959	223·5	2·2
1960	311·2	1·6
1961	320·3	1·5
1962	213·8	2·0
1963	179·5	2·5

(*Statistics on Incomes, Prices, Employment and Production*, March 65; *Annual Abstract of Statistics* 1964)

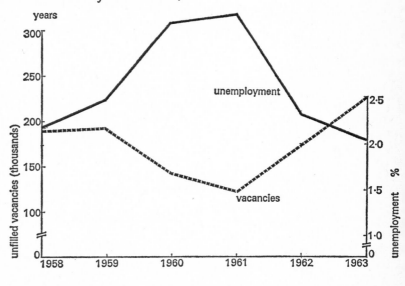

Fig. 22

There are two alternative solutions. We could make use of a double vertical axis (fig. 22), although there is obviously no natural relationship between the units of measurement and no comparison is therefore possible between the amplitude of the fluctuation of the two series. A comparison of turning-points in the two series can nevertheless be made, and we find that unfilled vacancies and unemployment move more or less together. As might be expected, unemployment is low when vacancies are high and vice versa.

The second solution is to convert both series from absolute units of measurement to relative units of measurement. This can be done by using index numbers:

| Coal production | | | Wage-earners in industry | |
Year	Tons (millions)	Index	Number (000's)	Index
1952	210·8	100·0	706·2	100·0
1954	211·5	100·3	701·8	99·4
1956	207·4	98·4	697·4	98·8
1958	198·8	94·3	692·7	98·1
1960	183·9	87·2	602·1	85·3
1962	176·8	83·9	531·0	75·2

Here we have calculated the percentage change in each year of the series as compared with a fixed period (1952) Thus the original units of

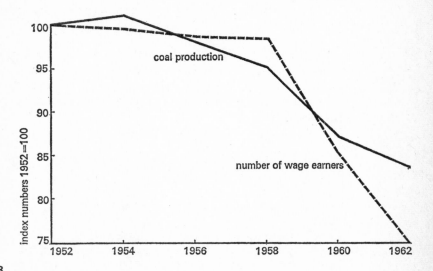

Fig. 23

measurement are completely discarded and both series have the same vertical axis (fig. 23).

The reader should notice one possible limitation to this second type of comparison. The selection of a different base year may completely alter the visual impression of the situation. If the percentage changes in the variables had been compared with 1962 instead of 1952, the series plotted in fig. 24 will result: the relationship between the variables seems to have altered. A study of the turning-points shows this to be untrue.

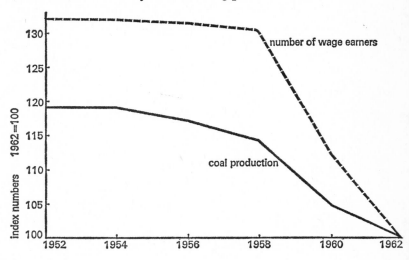

Fig. 24

2.5.5   Let us now return briefly to the question of absolute and proportionate rates of change. We have seen in 2.5.4 that a natural scale presentation of a variable indicates only absolute changes. If, however, we show not the actual values in the series but their logarithms, then proportionate changes are indicated:

Year	Sales of firm A (£000's)	Log sales
1958	80·0	1·9031
1959	120·0	2·0792
1960	180·0	2·2553
1961	270·0	2·4314
1962	405·0	2·6075
1963	607·5	2·7836

From an inspection of these hypothetical figures, it is possible to see that there has been a 50% increase in sales each year. In other words, there has been a constant proportionate rate of increase. This will always be shown by a straight line (as in fig. 25) when plotted on ratio or logarithmic scale.

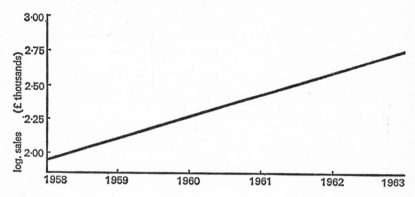

Fig. 25

The reason for this is that increasing the characteristic of the logarithm of any number by one unit results in a tenfold increase in the number.

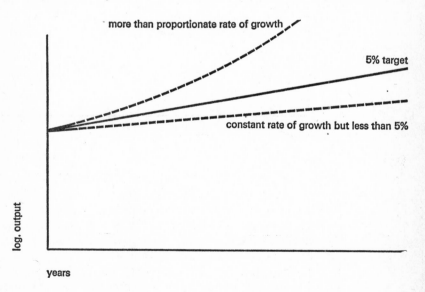

Fig. 26

63 The arrangement and presentation of data

For instance:

>0·6990 is the log of 5
>1·6990 is the log of 50
>2·6990 is the log of 500.

In general, we can say that any equal increases in a set of logarithms indicates an equal proportionate change in the numbers which they represent. From this it follows that equal vertical distances on the logarithmic axis of a graph also show equal proportionate increases in the variable.

The repercussions of these facts are of considerable value. A firm planning a 5% increase in output can see at a glance whether this is being achieved (fig. 26).

Similarly, because lines of equal slope must show equal proportionate rates of change, a firm can immediately evaluate the performance of different sales regions (fig. 27).

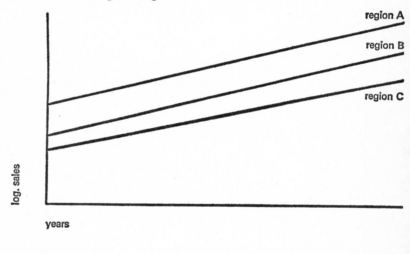

Fig. 27

While sales regions A and B are maintaining a 15% increase in sales per annum, region C is only managing a 10% annual increase. This may require further investigation and possibly remedial action.

While there are many advantages accruing from the use of semi-logarithmic (natural time scale and log variable scale) graphs, there are also certain disadvantages. Zero cannot be shown as it possesses no logarithm, and for the same reason negative values can never be taken into account. Even so, the modern preoccupation with growth rates of the economy, output and population demonstrates the importance of this method of presentation. The analysis of time series data is developed in greater detail in Chapter 6 and in *Statistics for the Social Scientist: 2 Applied Statistics*.

2.5.6 Cumulative graphs of time series

There are many occasions when data to be compared form constituents of the same aggregate or total. In these circumstances, it can be very useful to prepare cumulative tables from each year, and then plot these figures over the time period (fig. 28).

U.K. current expenditure on the social services (£ million)

	1954–5	Cumu-lative total	1956–7	Cumu-lative total	1958–9	Cumu-lative total	1960–1	Cumu-lative total
Benefits and assistance	909·9	909·9	1063·1	1063·1	1383·7	1383·7	1494·7	1494·7
Health and welfare	631·0	1540·9	750·4	1813·5	825·1	2208·8	982·3	2477·0
Education	432·0	1972·9	546·4	2359·9	669·9	2878·7	811·7	3288·7
Housing	103·9	2076·8	103·8	2463·7	110·7	2989·4	121·5	3410·2
		2076·8		2463·7		2989·4		3410·2

(*Annual Abstract of Statistics* 1963)

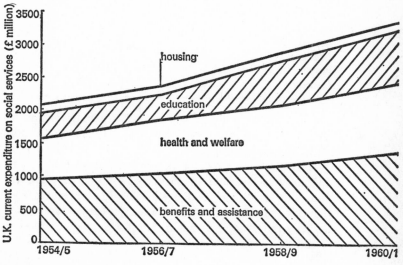

Fig. 28   Cumulative graph.

This presentation provides immediate evidence of the relative importance of the various constituents and the total expenditure on the social services, although it is not easy to analyse the changes in the various components. The changing base of the upper bands in the graph is the cause of this difficulty.

2.5.7 Z charts

2.5.7.1 As the final method of graphing variable data over time, let us consider the Z chart. This provides a running picture of the sales performance of any firm. Using the hypothetical data below let us produce a Z chart for the 1966 sales of a chain of men's wear retailers.

Month	Sales (£000's) 1965	1966	Cumulative sales 1966	Moving annual sales
January	10·8	12·3	12·3	134·7
February	8·7	8·5	20·8	134·5
March	9·2	9·4	30·2	134·7
April	10·3	11·0	41·2	135·4
May	11·6	12·5	53·7	136·3
June	10·8	12·1	65·8	137·6
July	11·4	10·9	76·7	137·1
August	10·1	10·3	87·0	137·3
September	10·1	10·7	97·7	137·9
October	11·1	11·0	108·7	137·8
November	11·9	12·3	121·0	138·2
December	17·2	18·7	139·7	139·7

The cumulative sales for 1966 are self-explanatory. The moving annual sales figure is found by summing the sales in the 12 months inclusive up to the month in question: i.e. the moving annual sales for January 1966 is the sales figure for January 1966 plus the sales recorded from February to December 1965, whilst for February 1966 it is February 1966 plus March 1965 to January 1966. In other words, the new month's figure is added in each case and the sales for the same month in the year earlier subtracted, e.g. the M.A.S. for September 1966 = 137·3 + 10·7 − 10·1 = 137·9.

The actual sales, cumulative sales and moving annual sales are now plotted on one graph and it will be seen (fig. 29) that they describe roughly the form of a letter Z. All three curves provide useful information. The actual sales show the month to month fluctuations, the cumulative sales show whether the increase in sales throughout the year is steady or whether some months are more erratic than others, and the moving annual sales give an impression of the long-term trend (whether there is a general increase in sales over time).

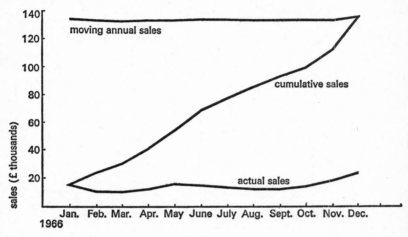

Fig. 29   Z chart.

2.5.7.2   At this point, we should examine some other presentational methods which are available. These are designed specifically for visual impact purposes, and as such are often used without any accompanying table. These are widely used in communicating numerical data to the general public and many devices are used to catch the eye of the viewer or reader; bright colours and shading can be employed with great effect. Once the attention has been gained, the purpose of the presentation is, in general, to show differences or changes in situations. One is more concerned with relative than with absolute magnitudes, so the way of bringing out this feature is the principal consideration.

2.6   Bar charts, pie charts and pictograms

2.6.1   A bar chart makes use of blocks of different heights in the same way as the histogram. The blocks in this case are always of equal base width so that the height is proportionate to the magnitudes represented; the basic principle is illustrated below in a compound bar chart (fig. 30).

Increase in U.K. total domestic income (percentage compared with 1938)

Income from	1938	1950	1962
Employment	100	250	568
Profits	100	230	470
Self-employment	100	218	338

(*National Income and Expenditure* 1965)

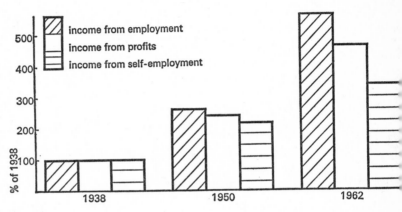

Fig. 30 Bar chart.

Notice that shading has been used with a key for more rapid under-standing. Numerical values might also have been included above each block for the same reason, thus enabling the scale of measurement on the left to be omitted. The same order of components has been main-tained throughout, but it should be remembered that a horizontal rather than a vertical presentation may sometimes be preferable.

2.6.2   Component bar charts

While the simple bar chart shown in fig. 30 is useful where one requires to compare the increases of the individual items, there is no way of gaining an immediate visual impression of the overall (total) change in income. This can be remedied by producing a component bar chart (fig. 31).

Assets of London clearing banks (£ millions)

Month	Liquid assets	Advances	Invest- ments	Special deposits
September 1962	2,559	3,428	1,234	151
March 1963	2,347	3,839	1,234	—
December 1963	2,723	3,961	1,281	—

(*Barclays Bank Review*, August 1964)

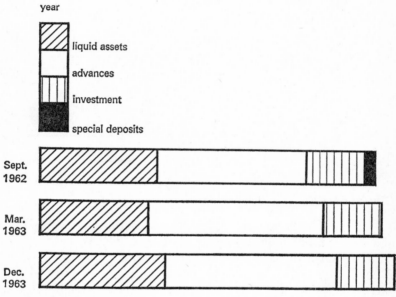

Fig. 31   Component bar chart.

Unfortunately it is no longer so easy to compare the components: a gain in one direction has produced a loss in another.

2.6.3   *Percentage component bar charts*

A third version which we must consider is the percentage component bar chart. This provides a visual impression of the relative importance of the different components (fig. 32).

69   The arrangement and presentation of data

Population distribution (percentage)

Age	Scotland	Northern Ireland	London
0–14	24·6	28·6	20·8
15–65	65·3	62·5	67·8
Over 65	10·1	8·9	11·3
Total (thousands)	5,114	1,373	10,928

(*Regional Statistics*, 1965)

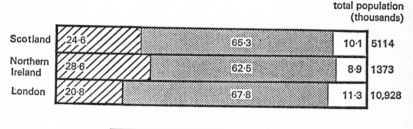

Fig. 32 Percentage component bar chart.

It may be advisable in these circumstances to write the actual percentages in the appropriate section of the bars, and to give the absolute totals to which these percentages refer.

2.6.4 *Pie charts*

Another method of representing variable quantities is given by the pie chart. A circle is divided into segments, the areas of which are proportionate to the values in question; this implies, of course, that the angles of each segment are proportionate in the same way (fig. 33).

There are certain objections to this form of presentation. Firstly, the eye finds considerable difficulty in distinguishing differences in angles; bar charts involving lengths are therefore preferable from this standpoint. Secondly, if one deals with sets of data in different places at different times and the aggregate of the individual components is increasing, then the overall size of the circle must increase; the calculation of these varying sizes is a somewhat tedious process, which should be avoided if possible, and again the bar chart is free from this criticism. It does seem, therefore, that only percentages can be shown, for they require a circle of consistent size.

United Kingdom imports, 1964 (£ million)

Source	Value	Percentage	Angle (degrees)
Overseas Sterling Area	1,631	32·6	117
Western Europe	1,628	32·5	117
North America	997	19·9	72
Latin America	260	5·2	19
Others	489	9·8	35
	5,005	100·0	360

(*U.K. Balance of Payments* 1965)

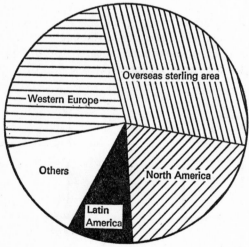

Fig. 33   Pie chart.

2.6.5   *Pictograms*

2.6.5.1   The final form of purely visual data presentation that is in wide currency is the pictogram. Small pictures of some object, conveying the variable involved, are employed: pictures of men can be used in population data, crates in import data, houses in construction data and so forth. In fig. 34, oil drums have been employed.

Fig. 34   Pictogram: import of oil to the United Kingdom.

It can be seen that in 1930 approximately $7\frac{1}{2}$ thousand million gallons of oil were imported; in 1950, 11 thousand million and in 1960, $12\frac{1}{4}$ thousand million.

The use of many pictures of the same size is far preferable to the drawing of a small number of pictures of different sizes. In the second case, either lengths, areas or volumes must be proportionate to the variable figures, for the reasons demonstrated in fig. 35.

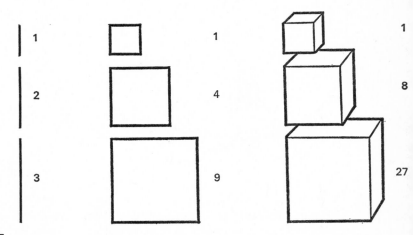

Fig. 35

One may easily change the proportionate relationship by moving from one to two or three dimensional pictures. Inadvertently the visual impression has been changed, although the basic unit of measurement is maintained.

2.6.5.2   There are no hard and fast rules concerning the construction of diagrams of the bar chart, pie chart, or pictogram type. There are many variants on the themes which have been mentioned, but all of them strive to get over a numerical message clearly and concisely. Anything which aids this objective is valuable and should therefore be included. The reader is strongly advised to look through government publications, bank reviews etc. where he will undoubtedly find many illustrations of these techniques being used in a very practical manner.

2.7   Correlation and contingency tables

2.7.1   Both correlation and contingency tables study relationships between variables; quantitative in correlation tables and qualitative in contingency tables. For example, we might look at income per head in a number of countries and the percentage of the population over the age of 18 receiving full-time education. We could construct a table of this bivariate data in the following form:

Country	Income per head (dollars)	Over 18s receiving full-time education (percentage of total population)
A	1,500	7·3
B	455	2·4
C	1,235	10·1
D	281	0·9
E	41	0·3
F	879	4·3
G	2,000	9·2
H	635	6·2

The word bivariate is used to indicate that for each country two variables are observed, so that it is possible to evaluate any connexion between them. The same thing applies to the qualitative data below, collected in a market research investigation.

73   The arrangement and presentation of data

Attitude to product X	North	Area of country Midlands	South
Like	51	23	35
Indifferent	28	65	39
Dislike	18	41	46

Here we show the attitude of consumers living in different parts of the country to a new product. The numerical values show the number of responses of each type in each region. This form of tabulation can be of great value to the firm in formulating its sales promotion policy. Such questions as where to advertise and how much to spend on advertising can be answered by reference to the sales resistance indicated in this table.

2.8 Examples

2.8.1 Frequency distributions

(a) A large car distributor markets motor vehicles for three manufacturers, X, Y, Z. For X and Y he sells saloons, shooting brakes and sports cars, while for Z only shooting brakes and commercial vehicles. The distributor requires an analysis of his sales over one year, broken down into quarters. He is interested in both the actual turnover figures and the relative position of the three manufacturers in these sales. Finally, he would like to know the percentage of the total sales of each manufacturer's vehicles attributable to the various types (i.e. saloons, shooting brakes etc.).

Draw up a skeleton table on which all this information may conveniently be shown.

(b) A sample of 100 patients' record cards shows the number of visits to the doctor's surgery in one year to be:

0	2	6	2	6	5	22	3	1	10
2	5	1	45	4	9	7	25	9	48
10	16	15	5	7	8	3	26	6	18
5	0	6	22	8	11	23	8	5	9
15	0	16	11	13	1	7	32	2	18
9	8	5	9	17	7	29	5	9	12
5	7	13	18	8	37	8	27	7	13
7	20	1	9	4	6	23	9	6	11
7	7	22	71	17	41	11	28	1	44
53	14	55	2	62	6	11	3	34	56

Group these figures into a frequency distribution using class intervals of suitable widths. Draw the histogram of your distribution and superimpose the frequency polygon. What information can be gained about the patients of this doctor?

(c) In a survey of families in a new town, one of the questions asked was 'How many domestic appliances do you own?' The question was confined to consumer durables. The fifty answers received are given below:

1	2	2	1	1
2	3	3	3	2
3	2	1	2	5
1	3	4	7	2
4	4	0	1	3
3	2	2	2	1
2	3	3	3	3
2	6	6	1	1
5	2	2	4	3
3	2	3	4	2

Construct a line chart from the above data and draw the 'more than' and 'less than' ogives.

(d) In a test run of cars of the same model supervised by a motoring association, the average mileage covered per gallon of fuel consumed was as follows:

18·6	33·4	25·3	27·8	30·6	31·9	33·0	26·3	24·9	29·4
20·0	26·2	28·1	33·1	37·5	22·5	39·1	32·9	33·8	52·6
32·5	34·6	32·7	9·5	38·5	29·6	25·3	49·5	30·1	27·9
26·9	23·8	36·0	38·0	27·5	32·3	34·2	23·1	34·7	29·0
34·1	38·6	25·9	40·6	58·3	29·3	36·8	27·1	34·9	34·0
29·1	38·3	35·2	28·9	28·0	33·7	23·8	30·6	31·6	21·8

A survey fifteen years earlier of a car of the same size showed the following results:

m.p.g.	Number of cars
Up to 19·9	14
20·0–24·9	28
25·0–29·9	70
30·0–34·9	38
35·0–39·9	12
40·0–59·9	8
	170

Using the same grouping, construct a frequency distribution for this data and compare the results of the two investigations by drawing the relative frequency polygons on the same graph.

(e) Construct relative cumulative frequency tables for the two sets of data in (d) and estimate the limits of mileage covered by the middle 50 per cent of vehicles in both cases. (Construct 'more than' cumulative frequency curves for this purpose.)

2.8.2 *Time series*

(a) Represent the following time series graphically. All necessary detail should be supplied; justify the method of presentation used.

Steam pressure in boiler subject to constant heat (i)

	lb. per cubic inch
Start up	0
+1 hour	2·51
+2 hours	6·31
+3 hours	15·85
+4 hours	39·81
+5 hours	100·00

Dividend paid by joint stock company (ii)

Year	Per cent
1950	4
1951	5
1952	4
1953	4
1954	4
1955	4
1956	$5\frac{1}{2}$
1957	6
1958	6
1959	6
1960	7
1961	$7\frac{1}{2}$
1962	6
1963	6
1964	6
1965	6

Taxes on expenditure in the United Kingdom (iii)

Year	£ million
1938	622
1946	1,573
1947	1,816
1948	2,013
1949	1,993
1950	2,065
1951	2,270
1952	2,292

(*National Income and Expenditure* 1965)

(b) For purposes of comparison, plot the two following sets of data on semi-log scale, using natural scale graph paper. Why are you asked to use semi-log graphs when this information could easily be plotted on natural scale?

77 The arrangement and presentation of data

Average weekly output in tons

Year	Plant A	Plant B
1956	68	272
1957	76	304
1958	54	216
1959	95	380
1960	83	332
1961	51	204

(c) What conclusion might be drawn from plotting the following series on a single graph? Use any appropriate method.

Index of retail prices (Jan. 1956 = 100)	Year	Average weekly earnings of manual workers £
109·0	1958	13·28½
109·6	1959	13·55½
110·7	1960	14·53½
114·5	1961	15·34
119·4	1962	15·86
121·7	1963	16·74½
125·7	1964	18·11

(*Monthly Digest of Statistics*, October 1965)

(d) Show this time series data in a cumulative form. What information is provided about the changing pattern of milk utilization?

The utilization of milk used for manufacture (million gallons)

Year to 31st March	Butter	Cheese	Condensed milk	All others
1955	56	126	73	56
1956	54	126	77	65
1957	99	196	102	74
1958	153	213	95	77
1959	81	161	85	87
1960	70	175	83	94

(*Monthly Digest of Statistics*, November 1961)

(e) The number of passenger-miles (in millions) covered by U.K. domestic airline services in 1962 and 1963 is as follows:

	1962	1963
January	29·2	32·2
February	29·8	33·1
March	39·9	48·1
April	52·8	63·2
May	57·6	71·5
June	86·5	94·1
July	101·6	109·4
August	101·5	113·9
September	84·3	92·9
October	51·0	61·8
November	38·5	46·0
December	33·7	47·9

(*Monthly Digest of Statistics,* October 1965)

Use this data to produce a Z chart for 1963. What emerges from this diagram?

2.8.3   *Charts*

(a) In 1962 EFTA countries took 14% of Britain's exports. This was divided up in the following manner:

	Per cent	Per cent increase in exports to each EFTA country
Austria	7	76
Sweden	29	52
Norway	14	52
Portugal	7	48
Switzerland	14	45
Finland	27	45
Denmark	22	18

Britain's exports to EFTA countries have been increasing rapidly during the past few years and between 1959 and 1963 the percentage rise

in the exports to the above are shown on the right. Thus with the ex
ception of Denmark the increase in our exports to these countries ha
been in excess of the average increase in our exports to the whole world
which between these years was 21%. Construct diagrams to demon
strate this information.

(b) Investment in Britain increased between 1951 and 1962 from £2,46
million to £4,390 million, representing 14·6% and 18·6% of Nationa
Product respectively. Of these totals, 24% went in 1962 to manufac
turing, 22% to distribution, 19% to dwellings, 12% to public services
11% to public utilities, 10% to transport and 2% to other purposes
Within these sectors of the economy, 37% of the investment was mad
in plant and machinery, 19% on dwellings, 11% on vehicles, ship
and aircraft, and 33% on other new buildings and works. Represen
this data diagrammatically.

(c) The World Bank's subscriptions and members are as follows:

Area	Subscription (£ million)	Subscription per head of population (£)
U.S.A.	2,268	11·9
E.E.C.	1,243	7·0
East Asia	1,079	1·2
U.K.	929	17·0
South America	414	2·7
EFTA	272	7·4
Canada	268	14·2
Africa	263	1·5
Australasia	250	18·5
West Asia	172	1·9
Rest of Europe	152	2·5
Central America	92	1·5
	7,402	

Show these two sets of figures diagrammatically to draw out the salien
features.

(d) The Census of Production shows the number of manufacturin
establishments in 1958 to be:

Conurbation	Establishments
Tyneside	1,112
West Yorkshire	6,301
Greater London	20,290
West Midlands	7,504
South-east Lancashire	6,857
Merseyside	1,882

(*Regional Statistics* 1965)

Show these figures in the form of a pictogram.

(e) In the years shown, Britain's imports (in £ million) have come from the following sources:

	1954	1957	1960	1962
Total	3359·2	4043·7	4540·7	4492·0
Sterling Area	1921·0	1512·5	1506·4	1500·7
North America	555·5	804·9	941·3	826·2
E.E.C.	390·6	419·0	661·1	708·5
EFTA	391·5	470·5	558·3	550·7
Rest of World		(Calculate these figures)		

(*U.K. Balance of Payments* 1965)

Show this information in the form of percentage component bar charts.

(f) In the first six months of 1964, the monthly average figures for exports to the main sources were as follows:

	£ million
Europe	138
Commonwealth	122
North and South America	46
Soviet block	8
Others	61
Total	375

(*U.K. Balance of Payments* 1965)

Construct a pie chart to represent these figures.

Chapter 3
The description of data: averages and dispersion

We saw in chapter 2 that it can be extremely valuable to set out the results of inquiries in a tabulated and diagrammatic form. The value of these methods lies in the ready understanding which they provide of the circumstances behind the data. We must now ask ourselves whether sufficient analytical detail is forthcoming from these visual presentations alone, and in general the answer must be in the negative.

Primarily we need some means of describing the situations with which we are confronted. A concise numerical description is often preferable to a lengthy tabulation, and if this form of description also enables us to form a mental image of the data and interpret its significance, so much the better. Fundamental to many techniques in statistics is the description of variable data; this chapter will therefore look at the meaning and calculation of 'averages' and 'measures of dispersion'.

3.1 Averages

3.1.1 Looking at the two frequency distributions shown below we should ask ourselves what exactly is the distinguishing feature.

| | Number of houses | |
Number of rooms	Suburb A	Suburb B
3 or 4	8	0
5 or 6	27	8
7 or 8	30	27
9 or 10	16	30
11 or 12	0	16

Inspection of the frequency polygons shown in fig. 36 gives the answer. We find that they have exactly the same shape. It is their position relative to the horizontal (variable) axis which distinguishes them.

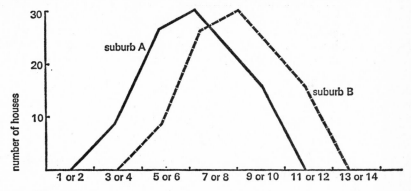

Fig. 36   Frequency polygons.

The polygon for Suburb B is further away from the origin than that for Suburb A. In fact the average number of rooms per house in B is 8·83 while in A it is 6·83. It would appear that 'measure of position' is synonymous to 'average'. Whichever terminology is preferred, we have here a means of positively analysing the difference between these two suburbs. It is apparent that there are larger houses in the second area than in the first, to the extent that there are on average two more rooms in each house.

3.1.2   In the strictly statistical sense the word 'average' is not self-explanatory and it will be noticed that the term appears between inverted commas in this discussion when it is employed in a technical context. This is because there are several types of average, each of which has a use in specifically defined circumstances. We shall now look at each of these in turn.

3.2   The arithmetic mean

3.2.1   The 'arithmetic mean' is the statistician's term for what the layman knows as the average. It can be thought of as that value of the variable series which is numerically most representative of the whole series. Certainly this is the most widely used average in statistics as applied to the social sciences. In addition it is probably the easiest to calculate, as is shown in this illustration:

Day	Receipts of a newsagent
Monday	£9·90
Tuesday	£7·75
Wednesday	£19·50
Thursday	£32·75
Friday	£63·75
Saturday	£75·50
Sunday	£50·70
Week Total	£259·85

The arithmetic mean sales per day in this week can be obtained by dividing the total receipts by 7 (the number of days).

The arithmetic mean $= \dfrac{£259 \cdot 85}{7} = £37 \cdot 12$ (to the nearest penny).

3.2.2   Using the conventions introduced in chapters 1 and 2, we denote the variable by x. The arithmetic mean is then the sum of the x's divided by the number of x's, namely n. The mathematical shorthand is therefore:

$$\frac{\Sigma x}{n} = \text{arithmetic mean}$$

As explained in 1.5.1 this should be $\dfrac{\sum\limits_{i=1}^{n} x_i}{n}$, which explicitly states the range of the variable over which the summation is to be taken. However, unless otherwise indicated this will always involve every value in the series.

Notice that the symbol $\bar{x}$ (pronounced bar-x or x-bar) is used to indicate the arithmetic mean.

3.2.3   The calculation of $\bar{x}$ when a frequency distribution has been formed is only a little more difficult.

Remembering that the (f) frequency column simply tells us how many students there are of each given age within the college, it would be possible to write down 17 forty-three times, 18 seventy-five times and so on. This would finally require the summation of 413 ages and division by 413. However, adding together forty-three 17's is the same as multiplying 43 by 17. Similarly $18+18+18+\ldots+18$ (75 times) is the same as 75×18. It seems to be expedient to multiply the number of students of each age by that age, giving fx. The summation of these products must produce the same answer as the sum of the original 413 ages.

Age of full-time students attending a college of technology x	Number of students f	fx
17	43	731
18	75	1,350
19	126	2,394
20	98	1,960
21	45	945
22	15	330
23	9	207
28	1	28
37	1	37
	413	7,982

The arithmetic mean is therefore given by

$$\frac{7982}{413} = 19\cdot33 \text{ years.}$$

In a general form this is

$$\bar{x} = \frac{\Sigma fx}{\Sigma f}$$

although in many cases Σf will be shown by N. The reason is firstly one of convenience and secondly one of consistency. The total number of observations in the series was represented before by n, and N has exactly the same implications although it has been found by summation rather than by simple counting.

3.2.4   The next difficulty which must be overcome is the calculation of the arithmetic mean when grouping has taken place. One is confronted with class intervals and obviously one cannot multiply the frequency value by the range of the variable shown. Some particular numerical value must be chosen which can be considered representative of the group. In the absence of information to the contrary, we shall work on the assumption that the observations in the group are evenly scattered between the two extremes of the class interval. This is an assumption which has been made in chapter 2, and which leads us to take the mid-point of the class interval as the representative value. Let us look at the stages in the calculation, using as an example the size of firms (based on the number of operatives employed) in the building industry in London.

Number of operatives employed	Number of firms in London f	Midpoints x	fx
none	4,682	0·0	0·0
1–10	6,485	5·5	35667·5
11–50	2,047	30·5	62433·5
51–100	286	75·5	21593·0
101–250	187	175·5	32818·5
251 and over	167	500·5	83583·5
	13,854		236096·0

Only one additional column of calculation is required: the midpoints, which assume the symbol x. This shows that the midpoints are being used to represent the variable range in each group. The arithmetic mean is given by

$$\frac{236096}{13854} = 17·04 \text{ operatives.}$$

The student will possibly be puzzled by the class interval headed 'none'. There is no conceptual difficulty involved here however, as we are looking at one-man businesses where the owner is the sole operative, no one else being employed by these entrepreneurs. Due weight is given to this type of business, which constitutes a third of all firms in the area, by the zero figure in the fx column. The group must not, under any circumstances, be omitted.

3.2.5   A much more basic and practical problem in the earlier examples was that of the magnitude of the arithmetic calculations. Obviously, these can be both tedious and time consuming, so a simplifying method would be a great asset in this direction. Such an approach does exist. It will be described as the 'assumed mean' method and can be illustrated in a simple form by considering the following series:

8, 8, 12, 16, 20.

The arithmetic mean is given by $\dfrac{8+8+12+16+20}{5} = 12·8.$

Let us suppose, however, that we had guessed at the value of the arithmetic mean (assumed the mean) from inspection and had decided that 14 was a likely figure. Now the deviations (d) of the original values (x) in the series round this assumed mean (a) are given by

$$(x-a) = d$$

$$\begin{array}{r} -6 \\ -6 \\ -2 \\ +2 \\ +6 \\ \hline -6 \end{array}$$

and the arithmetic mean of these deviations is

$$\frac{\Sigma d}{n} = \frac{-6}{5} = -1 \cdot 2.$$

It can be seen that the addition of this figure to the assumed mean gives the true mean, i.e. $a + \dfrac{\Sigma d}{n} = \bar{x}$ so that $14 + (-1 \cdot 2) = 12 \cdot 8$, which is correct. What we have achieved is the reduction of the original summation of numbers ranging from 8 to 20, to the summation of numbers with an absolute range of 2 to 6. The simplification process could be taken even further. We find that each deviation has a common factor (c) of 2. If this is taken out by division, then the values in the series of deviations is even more rapidly summed, the numbers lying between 1 and 3:

$$\frac{x-a}{c} = \frac{d}{c}$$

$$\begin{array}{r} -3 \\ -3 \\ -1 \\ +1 \\ +3 \\ \hline -3 \end{array}$$

Now the arithmetic mean of these reduced deviations, multiplied by the constant c and added to the assumed mean, produces the true mean:

$$a + \left(c \times \frac{\sum \frac{d}{c}}{n} \right) = \bar{x} \text{ so that } 14 + \left(2 \times -\frac{3}{5} \right) = 12 \cdot 8$$

3.2.6   It is evident that this procedure would be unnecessary in a simple series, but it is relevant in frequency distribution problems, where the numerical data is of such large proportions that the multiplication and summation is laborious. This is the case in the data below, showing the number of hours' overtime worked per week by a group of maintenance engineers in a large food-packing factory ($a = 6.25$, $c = 1.25$).

Number of hours' overtime	Number of men f	Midpoints x	d	$\dfrac{d}{c}$	$f\dfrac{d}{c}$
0 but less than 2·5	31	1·25	−5·00	−4	−124
2·5 but less than 5·0	48	3·75	−2·5	−2	−96
5·0 but less than 7·5	26	6·25	0	0	0
7·5 but less than 10·0	14	8·75	+2·5	+2	28
10·0 but less than 15·0	8	12·5	+6·25	+5	40
15·0 but less than 20·0	3	17·5	+11·25	+9	27
	130				−125

The average number of hours' overtime worked is given by

$$6.25 + \left(1.25 \times -\frac{125}{130}\right) = 5.05 \text{ hours}$$

There is very little difference from the technique set out in 3.2.5, except that we have taken account of the number of occurrences of each value of the variable. The stages in this calculation may be summarized as follows:

1. Find the midpoints of the class intervals.
2. Select one of the midpoints as the assumed mean and subtract it from all the others. The selection should be as close as possible to the true mean, although the final answer will in no way be affected by an inappropriate choice. In general, a midpoint near the middle of the distribution will be suitable.
3. Divide the resulting deviations round the assumed mean by any convenient constant.
4. Finally, multiply the reduced deviations by the appropriate frequencies and sum the products.
5. Now substitute the value of $\sum f\dfrac{d}{c}$ and Σf in the formula

$$\bar{x} = a + \left(c \times \frac{\sum f\dfrac{d}{c}}{N}\right)$$

3.2.7   It can be seen that we have developed here a technique of considerable importance, in which most of the arithmetic has been completed without resorting to the use of logarithms, thus achieving an overall reduction in computational time. In the context of the arithmetic mean calculation, however, there are several points to bear in mind. Firstly, the assumed mean method can also be used in a non-grouped distribution, in which case one of the actual values in the series (rather than a midpoint) is subtracted from all the others. Secondly, it should be an invariable rule to use this approach when the class intervals in the distribution are of the same numerical width, or are increasing by some multiple of the original width. Where the class intervals are unequal and not numerically related, the assumed mean method may prove no more advantageous than the straight $\frac{\Sigma f x}{N}$ procedure. The decision as to which method should be employed will be taken with reference to the overall structure of the distribution, but rests largely upon a little experience and practice.

3.2.8   Before concluding this discussion of the arithmetic mean, we should indicate some of the reasons for its predominant use as a measure of position or central tendency. It must be generally accepted that this is the best understood 'average' in statistics: it is relatively easy to calculate and in one sense it is a perfect average, for every value in the series is taken into account so that we obtain an exact answer which is suitable for further mathematical development. These are points on the credit side. There is one particular limitation to the use of the arithmetic mean which can prove to be a liability if not appreciated. We have stated that every value in a series is included in the calculation, whether it be high or low. Where there are a few very high or very low values, their effect can be to drag the arithmetic mean towards them. This may make the mean unrepresentative. For example, suppose one walks down the main street of a large city centre and counts the number of floors in each building. The following answers may be obtained:

5, 4, 3, 4, 5, 4, 3, 4, 5, 20, 5, 6, 32, 8, 27.

The arithmetic mean number of floors is $\frac{135}{15} = 9·00$, even though 12 out of 15 of the buildings have 6 floors or less. The three skyscraper blocks are having a disproportionate effect on the arithmetic mean. Some other average in this case would be more representative (see 3.4.1).

3.3 The geometric mean

3.3.1 There are other circumstances in which the arithmetic mean is discarded in favour of an alternative average. When relative changes in some variable quantity are averaged, for example, we prefer the *geometric mean*. Suppose it is discovered that a firm's turnover has increased during 4 years by the following amounts:

	Turnover	Percentage compared with year earlier
1958	£2,000	—
1959	£2,500	125
1960	£5,000	200
1961	£7,500	150
1962	£10,500	140

The yearly increase is shown in a percentage form in the right-hand column. The firm's owner may be interested in knowing his average rate of turnover growth. If the arithmetic mean is adopted he finds his answer to be

$$\frac{125+200+150+140}{4} = 153.75\%$$

In other words, the turnover in each of the years considered appears to be 53·75 per cent higher than in the previous year. If this percentage is used to calculate the turnover from 1958 to 1962 inclusive, we find that something is wrong:

$$153.75\% \text{ of } £2,000 = £3,075$$
$$153.75\% \text{ of } £3,075 = £4,728$$
$$153.75\% \text{ of } £4,728 = £7,269$$
$$153.75\% \text{ of } £7,269 = £11,176$$

It seems that both the individual figures and, more important, the total at the end of the period, are incorrect. Using the arithmetic mean has exaggerated the 'average' annual rate of increase in the turnover of this firm.

3.3.2 If the owner is honest with himself, he will want to rectify this false impression. The geometric mean will do just this. It is defined as being the nth root of the product of the values of the variable in question from the 1st to the nth:

$$\text{geometric mean} = \sqrt[n]{(x_1 \times x_2 \times x_3 \times \ldots \times x_n)}$$

In our situation we get

$$\sqrt[4]{(125 \times 200 \times 150 \times 140)} = \sqrt[4]{525000000}$$
$$= 151 \cdot 37\%$$

Now, if we use this value to obtain the individual turnover figures, we find:

151·37% of £2,000 = £3,027
151·37% of £3,027 = £4,583
151·37% of £4,583 = £6,937
151·37% of £6,937 = £10,500

Admittedly, the individual yearly turnover figures are incorrect, but this can be accepted. This would be similar to expecting a batsman with an average of 27·2 runs per innings to score this number of runs on each occasion that he bats. The final total, however, is completely accurate so that the owner of the firm is truly able to claim that on average each year's turnover is 51·37% higher than that in the previous year.

3.3.3   The reader should find no difficulty in calculating the geometric mean from a frequency distribution. The formula will be:

$$\text{geometric mean} = \sqrt[N]{(x_1{}^{f_1} \times x_2{}^{f_2} \times x_3{}^{f_3} \times \ldots \times x_n{}^{f_n})}$$

Each value of x thus has to be multiplied by itself f times, and the whole procedure becomes quite a formidable computational undertaking, if not an impossible one. If it were faced, then presumably the solution would be achieved by using log tables. The log of x_1 would be found and multiplied by f_1. The result would then be added to $f_2 \times \log x_2$, and so on. Finally, to take the Nth root, the sum of the frequencies times the logarithms of the variable values (or midpoints) will be divided by N. The antilog of this is the geometric mean required. Summarizing these steps gives:

$$\log \text{geometric mean} = \frac{\Sigma f \log x}{N}$$

This approach may also be used for a simple series such as our turnover percentages:

log 125 = 2·0969
log 200 = 2·3010
log 150 = 2·1761
log 140 = 2·1461
      4)8·7201
      = 2·1800

The antilog of 2·1800 is 151·4 which is very close to our earlier answer.

3.3.4　We should bear in mind that the geometric mean will be encountered less frequently than its arithmetic counterpart, and for this reason when the abbreviated term 'mean' is used in the text it will refer to the arithmetic mean only. Any reference to the geometric mean will be made quite explicit.

3.4　The median

3.4.1　Let us now return to the problem of the 'average' number of floors in the buildings at the centre of a city. We saw that the arithmetic mean was distorted towards the few extremely high values in this series and became unrepresentative. We could more appropriately and easily employ the *median* as the 'average' in these circumstances. The median is the middle value of the series when the variable values are placed in an order of magnitude. In a grouped frequency distribution, therefore, it is that value above and below which 50% of the observations lie. The median value can be ascertained by inspection in many series. For instance:

Height of buildings (*number of floors*)	*Retail price of motor-cars* (£) (*several makes and sizes*)
3	415
3	480
4	525　4 below
4　7 lower	608
4	719 = median price
4	1,090
5	2,059
5 = median height	4,000　4 above
5	6,000
5	
6	
8	
20	
27　7 higher	
32	

A slight complication arises when there are an even number of observations in the series, for now there are two middle values. The expedient of taking the arithmetic mean of the two is adopted as shown below.

Number of passengers travelling on a bus at six different times during the day

---

5
14
18
23 = median values
34
47

$$\text{median} = \frac{18+23}{2} = 20.5 \text{ passengers}$$

3.4.2   As stated on several earlier occasions, an ungrouped frequency distribution is no more than a concise representation of a simple series, so that the same approach as that discussed above would seem relevant.

Comprehensive school

Number of pupils per class	Number of classes	Cumulative frequency
23	1	1
24	0	1
25	1	2
26	3	5
27	6	11
28	9	20
29	8	28
30	10	38
31	7	45
	45	

There are 45 classes in all, so that we require as the median that class size below which there are 22 classes and above which there are 22 classes. In other words, we must find the 23rd class in an ordered list. We could simply count down noticing that there is one class of 23 children, 2 classes with up to 25 children, 5 classes with up to 26 children,

11 with up to 27 children, 20 with up to 28 children and 28 with up to
29 children. Evidently, the third of these last 8 classes of 29 children is
the median class size. It will not invariably be the case, however, that
the mental counting procedure will be simple. For this reason it is
preferable to construct a cumulative frequency table. From this we can
immediately discover that there are only 20 classes of 28 pupils or less,
while there are 28 of 29 or less. The median (23rd) must therefore be
one of the classes of 29 pupils. As a mechanical rule, that value which
has a cumulative frequency figure just in excess of the half-way position
is the median.

3.4.3   The only real difficulty in finding the median occurs when one is
presented with a frequency distribution with grouping. It has been stated
that here the median splits the observations into halves. This is shown
schematically in fig. 37.

Although the identification of the median group or class interval
presents little difficulty, to establish the position of the median within
this group is rather more troublesome:

Manufacturer's test

Breaking strength of steel cables	Number of cables	C.F.
Less than 8·0 tons	1	1
8·0 but less than 8·2 tons	7	8
8·2 but less than 8·4 tons	13	21
8·4 but less than 8·6 tons	23	44
8·6 but less than 8·8 tons	47	91
8·8 but less than 9·0 tons	39	130
9·0 but less than 9·4 tons	19	149
9·4 but less than 10·0 tons	5	154
10·0 tons and over	3	157
	$\overline{157}$	

As suggested by fig. 37, the median position is the $\frac{N}{2}$th position
$\left(\frac{157}{2} = 78\cdot5\text{th}\right)$ in the series. As before, we discover the median group by
looking for the next higher figure to 78·5 in the cumulative frequency
table. This is 91, so that the median lies somewhere between 8·6 and 8·8
tons. To determine just where, we must first remember our earlier stipu-
lation which assumed the observations in each group to be evenly

distributed between the two class limits. Secondly we ask how many observations there are, short of the median (half-way) position, at the lower boundary of the median class interval. This number divided by the frequency of the median class interval establishes the fraction of the numerical width of this class interval which must be added to the lower boundary to arrive at the median itself.

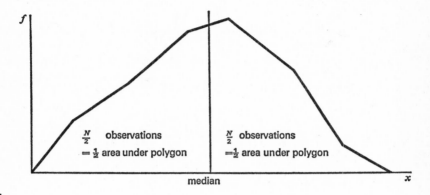

Fig. 37

In the context of the above distribution, we find that there are 44 cables which broke at a strain of less than 8·6 tons. At 8·6 tons therefore we are short of the median position by $78·5 - 44 = 34·5$ cables. The 34·5 cable out of the 47 which failed at 8·6 to 8·8 tons stress is the one we want. $\frac{34·5}{47}$ of 0·2 tons (the numerical width of the group) is 0·147 tons, which we add to 8·6 and arrive at 8·747 tons as the median value.

We can summarize these steps in the following formula:

$$\text{median} = L_m + C_m \times \left(\frac{\frac{1}{2}N - F_{m-1}}{f_m} \right)$$

Here L_m is the lower boundary of the median group, C_m is the class width of the median group, and f_m is the frequency of the median group. N is as usual the total number of observations in the distribution, and finally F_{m-1} is the cumulative frequency figure corresponding to the group preceding the median group. In this particular example, the formula yields:

$$8·6 + 0·2 \times \left(\frac{78·5 - 44}{47} \right) = 8·747 \text{ tons}$$

3.4.4   As a minor but important aside, we should notice that in all median calculations from grouped distributions we assume the variable to be continuous. This may look like a contravention of the principles laid down earlier, when it was explicitly stated that the continuous variable is a useful but nevertheless theoretical concept. The justification for such an apparent contradiction can best be shown schematically in fig. 38. Suppose we try to represent the following distribution diagrammatically:

Number of dwellings occupied since marriage	Number of families
1– 2	38
3– 4	41
5– 6	23
7–10	12
11–15	3
	117

The frequency heights present little difficulty. It is the plotting of the class intervals which proves to be a problem. If the original approach to a discrete series is adopted (2.4.2) we get the following form:

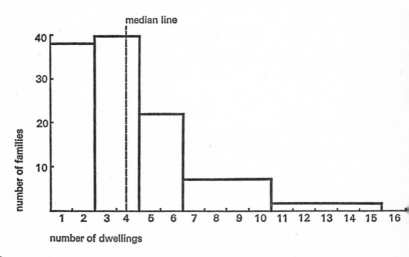

Fig. 38

A given length of the horizontal axis represents each value of the variable. Now if we superimpose a line on this histogram which equally divides the area in the blocks, it falls in the middle of the 4-dwellings value. Can we therefore suggest that 4 dwellings is the median? It would seem dubious to do so, for had the line been a little to the left or a little to the right, we would still get 4 as the answer. There is obviously only one median value, so that there would seem to be something wrong with this method of presentation.

Let us try showing one point on the horizontal axis for each variable value, so that the histogram now looks like this:

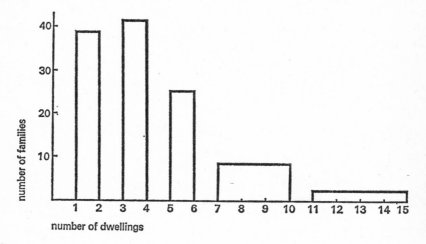

Fig. 39

Here there are parts of the axis which have no value at all, and the histogram blocks are detached. If we draw the blocks together, then a particular point has two values: one point will show 2 and 3, another 4 and 5, a third 6 and 7 etc. This also is most unsatisfactory.

A compromise between these possibilities produces the solution. If we convert the discrete series into a continuous one by making the class boundary the half-way position between the end of the one original class interval and the start of the next, we have a continuous scale of measurement with the histogram blocks adjacent (fig. 40).

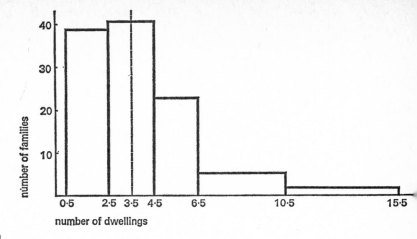

number of dwellings

Fig. 40

The line splitting the area under the histogram blocks now intersects the horizontal axis at a specific measurement, namely 3·5 dwellings. This is the answer obtainable from the formula method:

$$2\cdot5+\left(2\times\frac{20\cdot5}{41}\right) = 3\cdot5 \text{ dwellings}$$

3.4.5 *Quartiles, deciles and percentiles*

The median divides the area under the frequency polygon into halves. A further split to produce quarters, tenths or hundredths is equally possible and extremely useful for analysis. We are often interested in the highest 10 per cent of some group or the middle 50 per cent of another.

The *quartiles*, together with the median, achieve the division into 4. The calculation of the first quartile (Q_1) and the third quartile (Q_3) follows that established for the median, except that the Q_1 position is given by $\frac{N}{4}$ and the Q_3 position by $\frac{3N}{4}$; the second quartile (Q_2) is synonymous with the median.

The first and third quartiles for the distribution shown in 3.4.3 (breaking strength of steel cables) is as follows:

$$Q_1 = 8\cdot4+0\cdot2\times\left(\frac{39\cdot25-21}{23}\right) = 8\cdot56 \text{ tons}$$

$$Q_3 = 8\cdot8+0\cdot2\times\left(\frac{117\cdot75-91}{39}\right) = 8\cdot94 \text{ tons}$$

The *deciles* and the *percentiles* give the division into 10 and 100 respectively and once more follow the established procedure. The decile and percentile positions are found from

$$\frac{N}{10}, \frac{2N}{10}, \frac{3N}{10} \cdots \frac{9N}{10} \text{ and } \frac{N}{100}, \frac{2N}{100}, \frac{3N}{100}, \frac{4N}{100} \cdots \frac{99N}{100}$$

3.4.6 *The median by interpolation*

3.4.6.1   In 2.4.4 we saw how to construct an ogive, or cumulative frequency, curve. We also observed that it could be useful for estimating the number of observations lying between any two values of the variable range. Specifically we noticed that at the intersection of the 'less than' and the 'more than' ogives we had a value above which and below which 50 per cent of the observations lay. In other words, we had estimated the median by interpolation from a graph.

In practice, it is quite unnecessary to calculate and draw both ogive curves; one is quite sufficient. If a horizontal line is drawn from the $\frac{N}{2}$th position on the cumulative frequency axis, a perpendicular from the point at which it cuts the axis readily identifies the median.

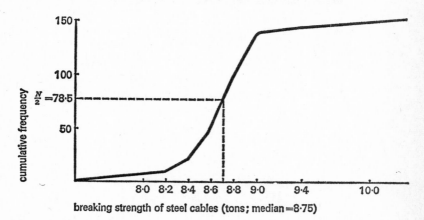

Fig. 41   Median by interpolation.

In a similar manner the quartiles, deciles and percentiles may be estimated from the ogive.

3.4.6.2   We have seen one reason for preferring the median as the 'average'. Besides overcoming the problem of a few extremely high or low figures in a series, it is also exceedingly valuable when one encounters an incomplete series having open-ended class intervals. Except in the unlikely event of the median falling within the open-ended group, there is no need to estimate the upper or lower boundary. Finally, we might notice at this point that the median is sometimes calculated as well as the arithmetic mean. The relationship between them (i.e. which is the higher) can throw much light upon the overall structure of the distribution (see 3.9.1).

3.5 The mode

3.5.1   If the manager of a men's clothing store were asked about the 'average' size of hats sold, he would probably think not of the arithmetic or geometric mean size, or indeed the median size; instead he will in all likelihood quote the specific size which is sold most often. He is using what the statistician describes as the *mode*, or that value in a series occurring most frequently. This 'average' is of far more use to him as a businessman. The modal size of all clothing is the size which he must stock in the greatest quantity and variety in comparison with other sizes. Indeed, in most inventory (stock level) problems one needs the mode more often than any other measure of position.

3.5.2   By definition, there can be no mode in a simple series, where no value occurs more than once. We must confine ourselves to frequency distributions. In the ungrouped distribution identification of the mode is immediate: one simply finds that value which has the highest frequency figure. For instance, an airline might find the number of passengers carried on a forty-seater plane in fifty flights to be:

Number of passengers	Number of flights
28	1
33	1
34	2
35	3
36	5
37	7
38	10
Mode = 39	13
40	8
	50

The mode is obviously 39 passengers, and the company is probably quite satisfied that a 40-seater is the correct-sized aircraft for this particular service.

3.5.3   When dealing with a grouped frequency distribution, the modal group is once more easily recognizable. As in the case of the median, the difficulty is that of deciding where within the modal group the mode lies. One might simply apply again the idea that the observations in this group are evenly distributed over the range of the class interval, and in consequence select the midpoint as the estimate of the mode. This is quite valid in an approximately symmetrical distribution, but for an asymmetrical distribution such as that shown below we might justifiably have reservations:

Size and number of orders received during one week by a wholesaler of electric power tools

Size of order (units)	Number of orders
Less than 10	15
10–19	23
20–29	47
30–39	42
40–49	31
50 and over	18
	176

The modal size can be seen to be between 20 and 29 power tools. However, as there are more orders of between 30 and 39 than of 10 to 19, it would seem that the maximum number of orders would be for 25 to 29 rather than 20 to 24 tools. This, incidentally, might well have shown up if a different grouping of the original data had been used (e.g. class intervals of 5 rather than 10).

It would seem quite valid from this discussion to suggest that the mode be drawn away from the midpoint of the group in which it lies, depending upon the number of observations in the preceding and following groups. This idea leads to the following formula:

$$\text{mode} = L_m + C_m \times \left\{ \frac{f_m - f_{m-1}}{2f_m - (f_{m-1} + f_{m+1})} \right\}$$

101   The description of data: averages and dispersion

where L_m and C_m are respectively the lower boundary and the class interval width of the modal group, while f_m, f_{m-1}, and f_{m+1} represent the frequencies of (respectively) the modal group, the group preceding and the group following it. If f_{m-1} is much larger than f_{m+1}, then the fraction to be multiplied by C_m will be small. If on the other hand f_{m+1} is the larger, then the fraction will be closer to unity.

In the example above the mode is given by:

$$19 \cdot 5 + 10 \times \left[\frac{47 - 23}{2 \times 47 - (23 + 42)} \right] = 27 \cdot 8 \text{ power tools}$$

(as in 3.4.4 the series is made continuous).

3.5.4   The major limitation of the procedure outlined in 3.5.3 is its arbitrary nature. We have chosen to take into account only the influence of the two class intervals adjacent to the modal group. We might easily have included the other frequency figures and thus have established an alternative formula (the reader may well encounter different formulae in other textbooks). It is this very weakness which produces a second limitation, namely that the mode does not lend itself easily to further mathematical treatment (see T. Dalenius in the *Journal of the Royal Statistical Society*, June 1965). Nevertheless, in the specific circumstances it is a valuable concept which is easily understood and ascertained from ungrouped series. Once again, like the median, it is not affected by very high or low values.

3.6   Scatter, spread or dispersion

3.6.1   Just as variable series differ with respect to their location on the horizontal axis (having different 'average' values), so they differ in terms of the amount of variability which they exhibit. In a technical college it may well be the case that the ages of a group of first-year students are quite consistent, e.g. 17, 18, 18, 19, 18, 19, 19, 18, 17, 18, 18 and 18 years. A class of evening students undertaking a course of study in their spare time may show just the opposite situation, e.g. 35, 23, 19, 48, 32, 24, 29, 37, 58, 18, 21 and 30.

We obviously need to be aware of the amount of dispersion or scatter in a series if we are to come to useful conclusions about the situation under review. This is perhaps best seen from studying the two frequency distributions below.

3.6.2   The sizes of the classes in two comprehensive schools in different areas are as follows:

Number of pupils	Number of classes Area A	Area B
10–14	0	5
15–19	3	8
20–24	13	10
25–29	24	12
30–34	17	14
35–39	3	5
40–44	0	3
44–49	0	3
	60	60

If the arithmetic mean size of class is calculated, we discover that the answer is identical: 27·33 pupils in both areas. Even though these two distributions share a common average, it can readily be seen that they are entirely different (fig. 42).

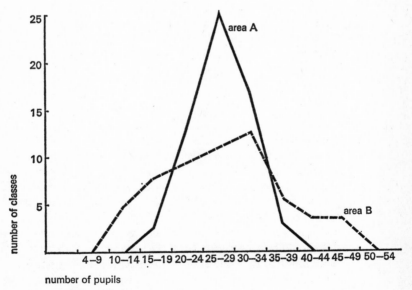

Fig. 42 Common means, different spread.

3.6.3   The question which must be posed and answered is 'How can these two situations be distinguished?' We need a measure of variability or

dispersion to accompany the relevant measure of position or 'averag
used. The word 'relevant' is important here for we shall find one measu
of dispersion which expresses the scatter of values round the arithmet
mean, another the scatter of values round the median.

3.7 The mean deviation and the standard deviation

How are we to decide upon the amount of dispersion round the aritl
metic mean? It would seem reasonable to compare each observed valu
in the series with the arithmetic mean of the series. Let us do this fo
the simple data below, which shows the number of fatalities in motorwa
accidents in one week.

Day	Number of fatalities x	Each value – arithmetic mean $x - \bar{x}$
Sunday	4	0
Monday	6	+2
Tuesday	2	−2
Wednesday	0	−4
Thursday	3	−1
Friday	5	+1
Saturday	8	+4
Total	28	0

The arithmetic mean number of fatalities per day is $\frac{28}{7} = 4$; this value
has been subtracted from each daily figure. The answers are therefore
negative when the daily figure is less than the mean (4 accidents) and
positive when the figure is higher than the mean. It does seem, however,
that our efforts have been in vain, for we find the total amount of dispersion as shown by summing the $(x - \bar{x})$ column to be zero. In fact
this should be no surprise, for if the arithmetic mean is to be numerically
representative then the sum of the positive and negative deviations of
the individual values round it should equate to produce a zero answer.

Our dilemma might at first sight seem insoluble, for by this criterion
it appears that no series has any dispersion. Yet this we know to be
nonsense, and in practice there are two possible approaches for overcoming the difficulty.

3.7.1 The mean deviation

Surely we might look at the numerical difference between the mean and the daily fatality figures without considering whether these are positive or negative? These absolute differences are given the symbol $|d|$, which is pronounced 'mod. d'.

x	$|d|$
4	0
6	2
2	2
0	4
3	1
5	1
8	4
	14

By ignoring the sign of the deviations we have achieved a non-zero sum in our second column. $\Sigma|d|$ might well be used as a measure of dispersion by itself, for in a different week we may find a very different figure:

x	$|d|$
4	0
5	1
4	0
3	1
6	2
2	2
4	0
28	6

$\Sigma|d|$ in this case equals 6, showing a very much smaller spread of values round an identical arithmetic mean figure.

The reader has perhaps realized that the use of the total of the differences suffers from a serious weakness. It is common sense to see that the magnitude of the $|d|$ column will depend as much upon the number of observations in the series as on the amount of variability. If we are to compare situations in which the number of observations is not the same, we need to average the differences. The average which is used is the arithmetic mean, and in consequence we call this measure of dispersion the *mean deviation* given by the formula

$$\frac{\Sigma|d|}{n}$$

The mean deviation number of fatalities is therefore $\frac{14}{7} = 2$.

The approach which we have adopted here is certainly simple and from the standpoint of computational speed, expedient. In problems involving descriptions and comparisons alone, the mean deviation can validly be applied; but because the negative signs have been discarded further theoretical development or application of the concept is difficult. We shall therefore generally prefer the *standard deviation* (shown by s) as the measure of dispersion associated with the mean.

3.7.2 *The standard deviation*

3.7.2.1 Let us now consider the second method of overcoming the zero sum, $\Sigma(x-\bar{x})$, which makes use of the fact that the square of both positive and negative numbers is positive. Squaring each deviation round the mean:

x	$(x - \bar{x})$	$(x - \bar{x})^2$
4	0	0
6	+2	4
2	−2	4
0	−4	16
3	−1	1
5	+1	1
8	+4	16
		42

Both $(-2)^2$ and $(2)^2$ equal 4, both $(-4)^2$ and $(4)^2$ equal 16, and both $(-1)^2$ and $(1)^2 = 1$. $\Sigma(x-\bar{x})^2 = 42$ is now positive and has been achieved without 'bending' the rules of mathematics. Rather than use the total, we prefer to average once again so that our measure of dispersion (the *variance*) is given by:

$$\frac{\Sigma(x-\bar{x})^2}{n}$$

In our example the answer is $\frac{42}{7} = 6$.

The variance is frequently employed in the statistical method, but it should be noted that the figure achieved is in 'squared' units of measurement. Our variance is 6 squared fatalities, which is obviously ridiculous. In order to reduce the variance to the original unit of measurement we usually take the positive square root. The result is described as the *standard deviation* with the formula

$$\sqrt{\frac{\Sigma(x-\bar{x})^2}{n}}$$

so that the standard deviation number of fatalities is $\sqrt{\frac{42}{7}} = 2\cdot45$ accidents.

3.7.2.2   In the practical situation of computing the standard deviation (or variance) it can be tedious to first ascertain the arithmetic mean of a series, then subtract it from each value of the variable in the series, and finally to square each deviation and sum. It is very much more straightforward to use the following identity (see also 3.10.6):

$$\sqrt{\frac{\Sigma(x-\bar{x})^2}{n}} \equiv \sqrt{\left\{\frac{\Sigma x^2}{n}-\left(\frac{\Sigma x}{n}\right)^2\right\}}$$

For substitution in the second formula we require only the aggregate of the series (Σx) and the aggregate of the squares of the individual values in the series (Σx^2). In other words, only two columns of figures are called for. The number of individual calculations is also considerably reduced, as shown below:

Number of industrial disputes in the motor industry resulting in stoppages (each month)

Month	x	x^2
January	2	4
February	1	1
March	8	64
April	6	36
May	4	16
June	0	0
July	2	4
August	3	9
September	5	25
October	1	1
November	7	49
December	1	1
	40	210

Therefore $s = \sqrt{\left\{\dfrac{210}{12} - \left(\dfrac{40}{12}\right)^2\right\}} = \sqrt{(17 \cdot 50' - 11 \cdot 11')}$

$$= \sqrt{6 \cdot 38'} = 2 \cdot 53 \text{ stoppages}$$

3.7.2.3   It should be evident that where one is calculating the standard devition from a frequency distribution, each squared deviation round mean must be multiplied by the appropriate frequency figure. Wheth x is an actual value in an ungrouped distribution or the midpoint one with grouping is immaterial, so we have:

$$s = \sqrt{\dfrac{\Sigma f(x - \bar{x})^2}{N}}$$

As before, this is exactly the same as

$$s = \sqrt{\left\{\dfrac{\Sigma f x^2}{N} - \left(\dfrac{\Sigma f x}{N}\right)^2\right\}}$$

which is again preferred from the computational standpoint. For stance, the standard deviation life of a batch of electric light bulbs wou be calculated as follows:

Hours life (hundreds)	Number of bulbs f	Midpoints x	fx	fx^2
0–5	4	2·5	10·0	25·0
5–10	9	7·5	67·5	506·25
10–20	38	15·0	570·0	8550·0
20–40	33	30·0	990·0	29700·0
40 and over	16	50·0	800·0	40000·0
	100		2437·5	78781·25

$$\text{Therefore } s = \sqrt{\left\{\dfrac{78781 \cdot 25}{100} - \left(\dfrac{2437 \cdot 5}{100}\right)^2\right\}}$$

$$= 13 \cdot 9 \text{ hundred hours or}$$
$$1390 \text{ hours}$$

3.7.2.4   In stating that each measure of dispersion is associated with a speci average, one implies that both the measure of dispersion and the measure of position to which it relates will be calculated in analysing a set of data. Now if the arithmetic mean has been found using the assumed mean or deviation method, it would seem necessary to develop an extension of the same method for ascertaining the standard deviation of the distribution. This, fortunately, is a very straightforward matter. Using the assumed mean method in the series below with $a = 7$, we find the following deviations (section 1):

Section 1		Section 2		Section 3	
x	$d=(x-a)$	x^2	d^2	$\dfrac{d}{2}$	$\left(\dfrac{d}{2}\right)^2$
5	-2	25	4	-1	1
3	-4	9	16	-2	4
1	-6	1	36	-3	9
7	0	49	0	0	0
9	$+2$	81	4	$+1$	1
5	-2	25	4	-1	1
11	$+4$	121	16	$+2$	4
13	$+6$	169	36	$+3$	9
54	-2	480	116	-1	29

What emerges from this is that although the origin of measurement has been changed from 0 to 7, the relationship between each pair of d's is identical to that between each pair of x's. The numerical difference between the largest and smallest value in each is 12, i.e. $13-1$ and $6-(-6)$; the difference between 5 and 3 is the same as that between -2 and -4; and so on. Indeed, the variability of the deviations is exactly the same as the variability in the original series. For this reason, we shall find that the standard deviations of the d's and the x's are identical. To show this we have calculated x^2 and d^2, and summed all four columns (section 2).

For the x's:

$$s = \sqrt{\left\{\frac{480}{8} - \left(\frac{54}{8}\right)^2\right\}} = \sqrt{14 \cdot 4375}$$

$$= 3 \cdot 80$$

For the d's:

$$s = \sqrt{\left\{\frac{116}{8} - \left(\frac{-2}{8}\right)^2\right\}} = \sqrt{14\cdot4375}$$
$$= 3\cdot80$$

From this demonstration we may conclude that it is perfectly correct t calculate the standard deviations of the d's having already used t assumed mean method in obtaining the mean:

$$\sqrt{\left\{\frac{\Sigma d^2}{n} - \left(\frac{\Sigma d}{n}\right)^2\right\}} \text{ gives the same answer as } \sqrt{\left\{\frac{\Sigma x^2}{n} - \left(\frac{\Sigma x}{n}\right)^2\right\}}$$

What happens when, to simplify the calculations still further, t deviations round the assumed mean have been divided by a constan This has been done in section 3 of the table, with $c = 2$. The $\frac{d}{c}$ a $\left(\frac{d}{c}\right)^2$ columns have been obtained and summed. The standard deviatio would seem to be:

$$s = \sqrt{\left\{\frac{29}{8} - \left(\frac{-1}{8}\right)^2\right\}} = \sqrt{3\cdot609375}$$
$$= 1\cdot90$$

This is half as big as the true standard deviation, which suggests that change in the unit of measurement achieved by division or multiplic tion should be accompanied by a correction to the final answer of t same magnitude. Thus where d is divided by c the standard deviation given by

$$c \times \sqrt{\left[\frac{\Sigma \left(\frac{d}{c}\right)^2}{n} - \left(\frac{\Sigma \frac{d}{c}}{n}\right)^2\right]}$$

3.7.2.5   Similar comments are applicable to the calculation of the standa deviation from a frequency distribution where the following formu applies:

$$c \times \sqrt{\left[\frac{\Sigma f\left(\frac{d}{c}\right)^2}{N} - \left(\frac{\Sigma f\frac{d}{c}}{N}\right)^2\right]}$$

We will briefly illustrate the use of this formula by looking at the fo lowing distribution of the number of occupants in a hostel for hom less families during one year ($a = 74\cdot5$, $c = 5$):

Number of occupants in hostel	Number of days	Mid-points	d	$\dfrac{d}{c}$	$f\dfrac{d}{c}$	$f\left(\dfrac{d}{c}\right)^2$
0–19	3	9·5	−65	−13	−39	507
20–39	10	29·5	−45	−9	−90	810
40–49	7	44·5	−30	−6	−42	252
50–59	9	54·5	−20	−4	−36	144
60–69	25	64·5	−10	−2	−50	100
70–79	40	74·5	0	0	0	0
80–89	79	84·5	+10	+2	158	316
90–99	192	94·5	+20	+4	768	3,072
	365				+669	5,201

$$\bar{x} = 74 \cdot 5 + \left(\frac{5 \times 669}{365}\right) = 83 \cdot 66 \text{ occupants}$$

$$s = 5\sqrt{\left\{\frac{5201}{365} - \left(\frac{669}{365}\right)^2\right\}} = 16 \cdot 50 \text{ occupants}$$

3.7.2.6   To end the discussion of the standard deviation, we should notice a number of important details which have been implied but not stated explicitly in passing. Firstly, it must be a definite rule to calculate the standard deviation using the identical approach applied to find the mean. If $\bar{x}$ has been found from $\Sigma fx / N$ then s should be calculated from

$$\sqrt{\left\{\frac{\Sigma fx^2}{N} - \left(\frac{\Sigma fx}{N}\right)^2\right\}}$$

Similarly for $a + c \times \dfrac{\sum f\dfrac{d}{c}}{N}$ and

$$c\sqrt{\left[\frac{\sum f\left(\dfrac{d}{c}\right)^2}{N} - \left(\frac{\sum f\dfrac{d}{c}}{N}\right)^2\right]}$$

Whichever approach has been adopted for finding the mean, only one further column of calculations, Σfx^2 or $\sum f\left(\dfrac{d}{c}\right)^2$, is required in order to substitute in the appropriate formula. This rule is therefore prescribed from a consideration of time alone.

Leading from this last point, it should be clearly understood that Σfx^2

or $\sum f\left(\dfrac{d}{c}\right)^2$ is the sum of the frequency figures times the variable val⬛

midpoint or deviation squared, i.e. the sum of $f{\times}x{\times}x$ or $f{\times}\dfrac{d}{c}{\times}$

It is not $\Sigma(fx)^2$ nor $(\Sigma fx)^2$. This type of error betrays a lack of und⬛ standing firstly of the principles underlying the standard deviation, a⬛ secondly of the meaning of the Σ (sigma) notation.

3.7.3   *The coefficient of variation*

The standard deviation is expressed in absolute terms and is given the same unit of measurement as the variable itself. There are occasio⬛ however, when this absolute measure of dispersion is inadequate an⬛ relative form becomes preferable, e.g. if a comparison between ⬛ variability of distributions with different variables is required, or compare distributions with the same variable but with very differ⬛ arithmetic means.

The relative measure in question is the *coefficient of variation*, giv⬛ by:

$$\frac{100s}{\bar{x}}$$

It simply expresses the standard deviation as a percentage of the ari⬛ metic mean. To illustrate its use, consider the following examples.

The mean weekly earnings of a skilled factory worker in the Uni⬛ Kingdom are £19·50 with a standard deviation £4, while for ⬛ American counterpart the figures are $75 and $28 respectively. It is ⬛ immediately apparent which country has the greater spread of earnin⬛ The coefficient of variation quickly provides the answer. For ⬛ United Kingdom it is $\dfrac{4}{19\cdot5}\times100 = 20\cdot5$ per cent, and for the U.S⬛

$\dfrac{28}{75}\times100 = 37\cdot3'$ per cent. The spread of earnings in the U.S.A. immediately seen to be greater, and the reasons for this could then sought.

The crop yield from 20 one-acre plots of wheatland cultivated ordinary methods averages 35 bushels with a standard deviation of bushels. The yield from similar land treated with a new fertilizer is bushels, also with a standard deviation of 10 bushels. The yield va⬛ bility may seem to be the same, but in fact it has improved in view of higher average to which it relates. Again, the coefficient of variation shows this:

$$\text{Untreated} \quad \frac{10}{35} \times 100 = 28 \cdot 57 \text{ per cent}$$

$$\text{Treated} \quad \frac{10}{58} \times 100 = 17 \cdot 24 \text{ per cent}$$

3.8 The quartile deviation and range

3.8.1 One measure of dispersion associated with the median is the *quartile deviation* (or *semi-interquartile range* as it is sometimes called); this is the arithmetic mean of the deviations of the first and third quartiles round the median, and is given by:

$$\frac{(M-Q_1)+(Q_3-M)}{2} = \frac{M-Q_1+Q_3-M}{2}$$

$$= \frac{Q_3-Q_1}{2}$$

The quartile deviation should always be employed to indicate dispersion when the median has been adopted as the most appropriate average. Although simple to compute, it is much less satisfactory as a measure of dispersion than the standard deviation because it takes into account the spread of only two values of the variable round the median and this gives no idea of the rest of the dispersion within the distribution.

3.8.2 The comparison of the dispersion in series follows the same lines as the procedure for the standard deviation. The larger the quartile deviation, the greater is the scatter of values within the series, as the share-holding structure of the following two companies shows:

	Company X	Company Y
1st quartile	60 shares	165 shares
Median	185 shares	185 shares
3rd quartile	270 shares	210 shares

The quartile deviation for company X is $\frac{270-60}{2} = 105$ shares; for company Y it is $\frac{210-165}{2} = 22 \cdot 5$ shares. We find that there is a considerable concentration of shareholders with about the average (median) number of shares in company Y. In X, on the other hand, there are approximately the same number of both small, medium and large share holdings.

3.8.3 *The range*

The final measure of dispersion which we should mention is the *range* the difference between the two extreme values of a series, i.e. $x_n - x_0$ where x_n represents the highest value and x_0 the lowest. Its calculation is a question of mental arithmetic. The simplicity of the concept does not necessarily invalidate it, but in general it gives no idea of the distribution of the observations between the two ends of the series. For this reason it is used principally as an ancillary aid in the description variable data, in conjunction with other measures of dispersion.

3.9 Skewness and kurtosis

3.9.1 By providing information about the location of a series and the dispersion within that series it might appear that we have achieved a perfectly adequate overall description of the data. This may be true, but one should bear in mind that a measure of dispersion indicates only the *variability* of values round the average. It gives no information about the *distribution* of values round the average. It is quite possible to have two series which are decidedly dissimilar and yet have exactly the same arithmetic mean and standard deviation:

Age of onset of nervous asthma in children (*to nearest year*)	Children of manual workers	Children of non-manual workers
0–2	3	3
3–5	9	12
6–8	18	9
9–11	18	27
12–14	9	6
15–17	3	3
	60	60

In both series the mean is 8·5 years and the standard deviation 3·61 years. Yet by inspecting the polygons in fig. 43 it can be seen that one is symmetrical while the other is asymmetrical (lopsided to the right).

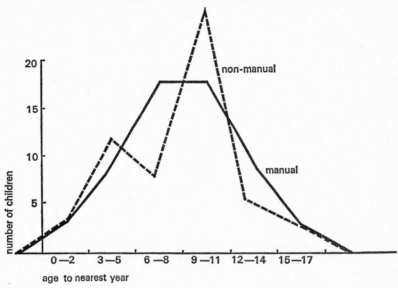

Fig. 43   Skewness.

3.9.2   The distinguishing feature here is the degree of asymmetry or *skewness* in the two polygons. To measure this feature, we use the fact that where there are extremely high or low values in a series the arithmetic mean will

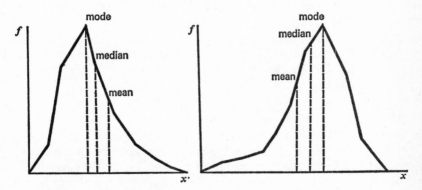

Fig. 44   Positive and negative skewness.

be drawn towards these more than the median, which in its turn
drawn towards them more than the mode. This is shown schematical
in fig. 44. Only in the case of a completely symmetrical distributio
will the mean, median and mode coincide (fig. 45).

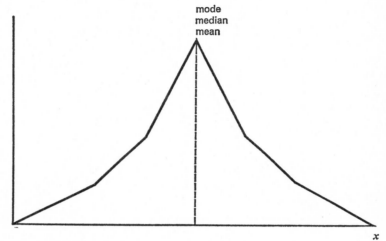

Fig. 45 A symmetrical distribution.

3.9.3   By subtracting one of these averages from one of the others, a positiv
negative or zero answer will result. The zero indicates perfect symmetr
The sign of a non-zero answer will distinguish the direction of the ske
ness, and the magnitude of the difference will indicate the degree
skewness.

Conventionally, the mode is subtracted from the mean and, to mak
the answer independent of the scale of measurement, the answer
divided by the standard deviation. *Pearson's coefficient of skewne*
results:

$$\frac{\text{mean}-\text{mode}}{\text{standard deviation}}$$

This formula has one disadvantage. We have already seen that th
calculation of the mode rests upon the most tenuous principles. Speci
cally, the mode is affected by the amount of skewness within the distr
bution and as such is best avoided when it is this very feature which
being estimated. To overcome this problem, we should notice that

moderately skewed distributions the median will be approximately twice as far from the mode as from the mean, i.e.:

$$\text{mode} - \text{median} = 2\ (\text{median} - \text{mean})$$

giving $\text{mode} = 3\ \text{median} - 2\ \text{mean}$.

Substitution in the original coefficient formula gives

$$\frac{\text{mean} - (3\ \text{median} - 2\ \text{mean})}{\text{standard deviation}} = \frac{3\ (\text{mean} - \text{median})}{\text{standard deviation}}$$

This derivation from the basic Pearson's formula is very much better, as the averages used are calculated from a sounder theoretical basis.

3.9.4   If the analysis of a series has been undertaken using the median and quartiles alone, then *Bowley's coefficient of skewness* (which requires no knowledge of the mean or standard deviation) is preferable. In an asymmetrical distribution the quartiles will not be equidistant from the median, and the amount by which each one deviates will give some indication of skewness. Where the lopsidedness is to the left, Q_1 will be closer to the median than Q_3; therefore by subtracting $2 \times$ median from Q_3 and Q_1 a positive answer will result. The opposite is true for skewness to the right (fig. 46).

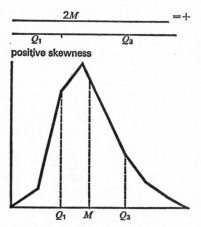
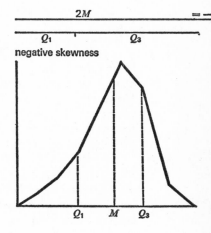

Fig. 46   Positive and negative skewness.

A relative measure is obtained in this case by dividing $Q_3 + Q_1 - 2M$ by the quartile deviation, so that Bowley's coefficient is given by:

$$\frac{Q_3 + Q_1 - 2M}{\frac{1}{2}(Q_3 - Q_1)} = \frac{2(Q_3 + Q_1 - 2M)}{Q_3 - Q_1}$$

3.9.5 Let us now calculate these various coefficients for the two distributions shown in 3.9.1.

	Children of manual workers	Children of non-manual workers
Mean	8·50 years	8·50 years
Standard deviation	3·61 years	3·61 years
Median	8·50 years	9·16' years
Q_1	6·00 years	5·50 years
Q_3	11·00 years	10·83' years
Quartile deviation	2·50 years	2·66' years

$$\text{Pearson's (modified)} = \overset{Manual}{\frac{3(8\cdot50-8\cdot50)}{3\cdot61}} \qquad \overset{Non\text{-}manual}{\frac{3(8\cdot50-9\cdot16')}{3\cdot61}}$$
$$= 0 \qquad\qquad\qquad = -0\cdot55$$

$$\text{Bowley's} = \frac{11\cdot00+6\cdot00-2\times8\cdot50}{2\cdot50} \qquad \frac{10\cdot83'+5\cdot50-2\times9\cdot16'}{2\cdot66'}$$
$$= 0 \qquad\qquad\qquad\qquad = -0\cdot75$$

It emerges that for a symmetrical distribution the coefficient w
always be zero, for a distribution skewed to the right (i.e. where t]
mean and median are drawn to the right of the mode) the answer w
always be positive, and for one skewed to the left always negativ
However, as is shown above, the magnitude of the positive or negati
coefficient will vary from one formula to another. In consequence o
formula should be used consistently, for the main use of the coefficie
is to evaluate changes in situations over time or space.

3.9.6 To complete the outline of the concept of skewness, it should
noticed that the measure which is widely accepted by mathematic
statisticians is the *third moment round the mean* (the first moment
zero; the second moment is the variance) divided by the cube of t
standard deviation which reduces it to a relative form, i.e.:

$$\frac{\Sigma(x-\bar{x})^3}{ns^3}$$

3.9.7 *Kurtosis*

This term may occasionally be encountered in statistical literature an
used to describe the amount of peakedness in a distribution, i

whether a distribution is very pointed with wide tails or humped with short tails. In the former case it is said to be *leptokurtic* (narrow humped) and in the latter *platykurtic* (broad humped). (See fig. 47.)

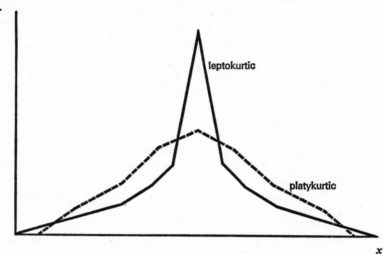

Fig. 47   Kurtosis.

The amount of kurtosis in a series is measured by the fourth moment round its mean divided by the fourth power of the standard deviation:

$$\frac{\Sigma(x-\bar{x})^4}{ns^4}$$

The numerical dividing line between a platy- and leptokurtic distribution is a value of 3; this is the coefficient which results from a 'normal' distribution. Any value greater than 3 indicates a leptokurtic distribution, while one of less than 3 shows platykurtosis. The coefficient of kurtosis is principally used as a simple means of ascertaining whether a distribution conforms to the 'normal' form or not. If it does, then the properties of this distribution may be used in analysing the data. (See *Statistics for the Social Scientist: 2 Applied Statistics*, Chapter 1 for a discussion of the normal distribution.)

3.10   Mathematical notes

3.10.1   In 2.3.1 and 2.4.3 we stressed that distributions involving continuous variables were theoretical concepts. Measurements are made in practice to the nearest appropriate unit, and effectively they are discrete. True continuous frequency distributions have to be expressed in the form of an algebraic equation, enabling calculations to be made using the differential and integral calculus.

For the reader interested in this side of the statistical method let us now show how the mean, median and mode are found. It is stressed that an elementary knowledge of the calculus is assumed here. The reader lacking such a knowledge may omit this section without any loss of continuity.

3.10.2   It has been observed earlier that a continuous distribution is such that the variable can assume every value between two limits, a and b. The frequency of any infinitesimally small interval of the variable range is indeterminate, so that the total frequency is infinite. Let us therefore look at these distributions in terms of their relative frequencies. The relative frequency of any infinitesimal interval is given by $f(x)dx$, where $f(x)$ is a continuous function of x and is described as the *relative frequency density*.

The continuous curve $y = f(x)$ is the relative frequency curve for the distribution, so that the relative frequency of any interval w to v is given by the area under the curve between these ordinates, i.e. $\int_{w}^{v} f(x)dx$

and the sum of the relative frequencies is unity, i.e. $\int_{a}^{b} f(x)dx = 1$.

3.10.3   The arithmetic mean is defined as the sum of the products of the midpoints of the infinitesimal intervals denoted by x, and the frequencies —in this case relative frequencies $f(x)dx$—divided by the sum of the relative frequencies which we have already seen to be equal to 1:

$$\bar{x} = \int_{a}^{b} xf(x)dx$$

The median is defined as the point p, such that a perpendicular drawn from it bisects the area under the frequency curve.

$$\int_{a}^{p} f(x)dx = \tfrac{1}{2} = \int_{p}^{b} f(x)dx$$

Finally the mode is that point at which the relative frequency is greatest. This is found when the first derivative $\dfrac{dy}{dx} = 0$, and the second derivative $\frac{d^2y}{dx^2}<0$. This is demonstrated by graphing the various functions (fig. 48).

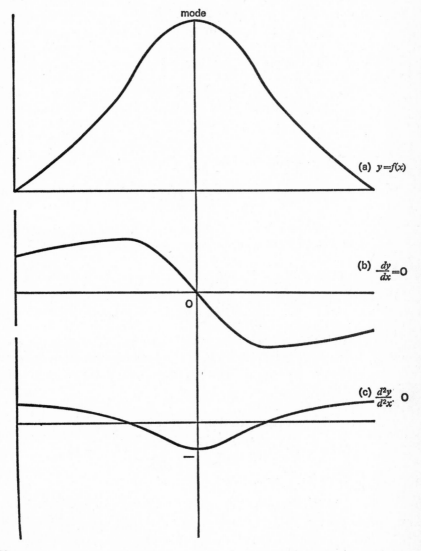

(a) $y=f(x)$

(b) $\frac{dy}{dx}=0$

(c) $\frac{d^2y}{d^2x}.$ 0

Fig. 48

3.10.4   These principles will now be applied to an actual example, that a continuous distribution with a relative frequency density given

$$f(x) = \frac{10x^2 - 2x^3}{104 \cdot 167} \text{ and a variable range from 0 to 5:}$$

$$\bar{x} = \int_0^5 x \left(\frac{10x^2 - 2x^3}{104 \cdot 167} \right) dx$$

$$= \frac{1}{104 \cdot 167} \left[\frac{10x^4}{4} - \frac{2x^5}{5} \right]_0^5$$

$$= \frac{1}{104 \cdot 167} \left[\frac{6250}{4} - \frac{6250}{5} \right]$$

$$= 3 \cdot 0$$

The median is found from $\int_0^p \left(\frac{10x^2 - 2x^3}{104 \cdot 167} \right) dx$

$$= \frac{1}{104 \cdot 167} \left[\frac{10x^3}{3} - \frac{2x^4}{4} \right]_0^p = \frac{1}{2}$$

p must therefore be such that $\dfrac{10x^3}{3} - \dfrac{2x^4}{4} = \dfrac{104 \cdot 167}{2} = 52 \cdot 083$

Thus $\dfrac{10p^3}{3} - \dfrac{2p^4}{4} = \dfrac{625}{12}$

Or $\dfrac{40p^3 - 6p^4 - 625}{12} = 0$

Using Newton's method of successive approximations we now ascerta the only root which lies in the range of the variable: 3·07. This is the fore the median; as a check notice that substitution in the final expressi above gives $\dfrac{1}{104 \cdot 167} (52 \cdot 083) = \dfrac{1}{2}$.

The mode is determined as follows:

With $y = \dfrac{10x^2 - 2x^3}{104 \cdot 167}, \dfrac{dy}{dx} = \dfrac{20x - 6x^2}{104 \cdot 167}$

Now for $\dfrac{dy}{dx} = 0$, $20x - 6x^2 = 0$

Factorizing gives $x(20 - 6x) = 0$, so that $x = 0$ or $\dfrac{20}{6} = 3 \cdot 33'$

But $\dfrac{d^2y}{dx^2} = 20 - 12x$, and for $\dfrac{d^2y}{dx^2} < 0$ x must be $3 \cdot 33'$

Therefore the mode is 3·33′.

3.10.5 These three results are shown in fig. 49 on the relative frequency curve in question. It is noticeable that negative skewness is exhibited and that the arrangement of the three averages confirms this.

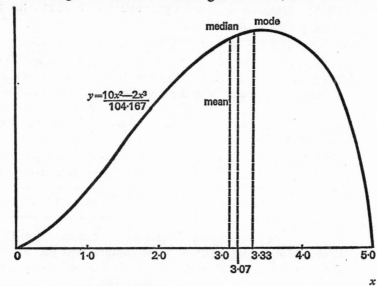

Fig. 49 Continuous frequency distribution.

3.10.6 The identity

$$\sqrt{\frac{\Sigma(x-\bar{x})^2}{n}} \equiv \sqrt{\left\{\frac{\Sigma x^2}{n}-\left(\frac{\Sigma x}{n}\right)^2\right\}}$$

is derived as follows:

$$\sqrt{\frac{\Sigma(x-\bar{x})^2}{n}} = \sqrt{\frac{\Sigma(x^2-2x\bar{x}+\bar{x}^2)}{n}}$$

$$= \sqrt{\frac{\Sigma x^2-2\bar{x}\Sigma x+n\bar{x}^2}{n}}$$

Putting $\bar{x} = \dfrac{\Sigma x}{n}$

$$= \sqrt{\left\{\frac{\Sigma x^2}{n}-2\frac{\Sigma x}{n}\frac{\Sigma x}{n}+\frac{n}{n}\left(\frac{\Sigma x}{n}\right)^2\right\}}$$

$$= \sqrt{\left\{\frac{\Sigma x^2}{n}-2\left(\frac{\Sigma x}{n}\right)^2+\left(\frac{\Sigma x}{n}\right)^2\right\}}$$

$$= \sqrt{\left\{\frac{\Sigma x^2}{n}-\left(\frac{\Sigma x}{n}\right)^2\right\}}$$

123 The description of data: averages and dispersion

3.11 Examples

Descriptive measures

(a) An inspection of samples of woollen cloth shows the following number of faults in each piece:

0, 1, 0, 2, 1, 0, 0, 2, 3, 5, 8, 4, 2, 1, 1, 0, 1, 1, 3, 4.

Calculate the arithmetic mean and standard deviation number of faults. These, together with the median, should enable useful conclusions to be drawn about the quality of the cloth. What are these conclusions?

(b) The residual error shown in the *National Income and Expenditure Blue Book* 1965 is as follows:

	1955	1956	1957	1958	1959	1960	1961	1962	1963	1964
£m	-45	178	275	287	202	-197	49	-71	55	-234

These figures represent the difference in the estimates of G.N.P. obtained from the aggregation of expenditures and the aggregation of factor incomes. Find the mean and standard deviation residual error for the years shown and comment on the result.

(c) A sample of small firms in manufacturing industry shows the following trade results to have been made in 1964–5:

Trade result	Number of firms
Loss	8
0–£250 profit	10
£250–£400 profit	18
£400–£550 profit	27
£550–£700 profit	25
£700–£1,000 profit	15
£1,000–£2,000 profit	5
£2,000 and over profit	2
	110

Find the median and quartile deviation. Would you expect the mean and mode to be higher or lower than the median? Explain your reasons.

(d) The expenditure on foreign travel in each year shown as a percentage of the previous year is:

Year	Percentage
1959	107·9
1960	113·4
1961	107·5
1962	105·0
1963	114·8
1964	108·3

What is the 'average' annual rate of growth of expenditure on foreign travel over these years?

(e) What are the mean, median and mode number of domestic appliances owned by the families shown in the data in 2.8.1, question (c)?

(f) A manufacturer receives components from two suppliers, A and B. A delivers in cartons of 20, B in cartons of 100. A sample of 100 cartons from each supplier is inspected in order to ascertain the quality of the components. The following results were obtained:

Supplier A Number below specified quality	Number of cartons	Supplier B Number below specified quality	Number of cartons
0	25	0–4	17
1	36	5–9	49
2	19	10–14	22
3	11	15–19	8
4	5	20–24	3
5	4	25–29	1

(i) What is the average proportion per carton of components below the specified quality, for the two suppliers?
(ii) Calculate the coefficient of variation for both series, in order to evaluate the amount of variation in quality from carton to carton.
(iii) From these results, will you renew your contract with both suppliers?

(g) Calculate the mode for both distributions in 2.8.1, question (
What can be concluded from your answers?

(h) The *Monthly Digest of Statistics* for February 1965 shows the a
distribution of the estimated total population of the United Kingdom
30 June 1964. This has been condensed and converted to a relati
frequency distribution:

Years of age	Percentage of total population
Under 10	16·00
10–19	15·00
20–39	25·52
40–59	26·12
60–79	15·40
80 and over	1·96
	100·00

Estimate, using any appropriate method:

(i) The percentage of the population entitled to vote (i.e. 21 years an
over).

(ii) The percentage of the population retired (65 and over) an
under 15.

(i) A firm wants to be 90 per cent certain that its two depots will no
run out of stock in any particular week. It finds from an investigation
that the weekly demand in the past five years at each of the depots ha
been as follows:

Number of units ordered	Number of weeks Depot I	Depot II
0–99	30	10
100–199	90	25
200–299	65	50
300–399	35	80
400–599	30	75
600 and over	10	20

Calculate the level of stock which should be held at the two depots, at the beginning of each week, if the chance of a run-out is to be kept at 10 per cent.

(j) A sample of 148 factories in a large industrial city were built in the following years:

	Number of factories
1850–1899	28
1900–1924	35
1925–1939	16
1945–1954	23
1955–1960	19
1961–1964	27
	148

What are the average and standard deviation age of these factories in 1965?

(k) Calculate Bowley's coefficient of skewness for the two distributions shown below. What do these coefficients indicate about the structure of trade unions over the ten-year period? Use any other measures necessary to comment on the situation.

Membership of trades unions	Number of trades unions	
	1953	1963
Under 100	142	115
100 and under 500	189	146
500 and under 1,000	72	57
1,000 and under 2,500	108	94
2,500 and under 5,000	73	58
5,000 and under 25,000	86	73
25,000 and under 50,000	17	18
50,000 and under 100,000	16	17
100,000 and over	17	18
	720	596

(*Annual Abstract of Statistics* 1965)

(l) Analyse the differences in the two industries below by making a necessary calculations.

Size of manufacturing establishments (June 1961)

Number of employees	Number of establishments	
	Vehicles	Food, drink and tobacco
11–24	388	1,425
25–99	810	2,359
100–499	437	1,224
500–999	120	153
1,000–1,999	82	85
2,000 and over	94	28
	1,931	5,274

(*Annual Abstract of Statistics* 1965)

Chapter 4
Index numbers in theory and practice

4.1 General principles

No text directed at the social scientist would be complete without a discussion of index number series. In regularly reporting and commenting on changes in index numbers relating to prices, production, imports and exports, wages etc., the press and other mass media have provided the general public with a superficial familiarity with this aspect of the statistical method. It is widely recognized that index numbers provide a measure of the relative change in some variable or group of variables at a specified date when compared with some fixed period in the past.

4.1.1   For instance, we may be examining the average price of houses in 1966 as a percentage of 1962:

	House prices	Index numbers (i.e. prices relative to 1962)
1962	£3,125	$\frac{3125}{3125} \times 100 = 100$
1966	£3,500	$\frac{3500}{3125} \times 100 = 112$

We would say that the *base year* in this case is 1962 (generally written as 1962 = 100) and that the index for 1966, the *current year*, is 112, showing there has been a 12 per cent increase in the average price of houses between 1962 and 1966.

4.1.2   Although many of the present-day economic indicators are produc in index number form, we have used a simple example involving pri changes because it was in connexion with the theory of value that t index number concept itself was developed. Economists since t beginning of the eighteenth century have been concerned with the pri level of goods and services, for the changing value of money has ma implications for economic policy decisions: inflationary tendenci need to be kept under control if the health of the economy and of t currency is to be maintained, as the collapse of the German econon under galloping inflation following the First World War readily bea witness.

Index numbers are also widely used by businessmen, who empl them to evaluate their trading positions in relation to competitors a rely on the national indices for wages, production, prices, sales, trar port charges and share prices to provide the simple background infc mation against which objective decisions may be taken. Labour leade must necessarily be interested in indices of earnings, wage rates, hou of work and retail prices (in so far as these reflect the cost of living), a many pay agreements are automatically tied to the cost of living, an appropriate index, so that the purchasing power of money income preserved at times of rising prices. Sociologists and educationi employ index number systems in the form of social class indices a intelligence quotients, which differ from index numbers measuri changes over time but nevertheless have a base or norm (equal to 1C to which reference is made in classifying individual people or famili

4.2 Weighted index numbers

4.2.1 The construction of an index series for one variable may be justifi from two interrelated standpoints. Often the significance of changes i tabulated list of figures (particularly if six or seven digits are involved) not immediately comprehensible. Changing from the absolute to t relative measure, facilitated by the use of indices, can largely remedy th In any case we are frequently more interested in a knowledge of p centage increases or decreases than in the absolute magnitude of chan so that even straightforward series may be converted with advantage in index number form.

4.2.2 Of far greater importance, however, is the conceptual problem volved in the construction of series which show in one figure the avera change in a group of associated variables compared with the base ye This may be explained by reference to a simple example. Suppose t a small manufacturer wishes to establish the changing picture of labour costs in his firm over the last ten years, the weekly wage rates paid in 1950, 1955 and 1960 being as follows:

Type of worker	1950	1955	1960
Unskilled	£8·75	£10·75	£14·50
Semi-skilled	£9·75	£12·50	£16·00
Skilled	£12·00	£14·00	£18·50
Clerical	£10·00	£12·25	£13·00

He may feel that some form of simple average of these wage rates will provide the basis for his index number calculations. For instance, he might work out the percentage wage for each category of worker for 1955 and 1960 as compared with 1950 and then average the indices obtained for each year (this we shall term the 'relatives' method). The main advantage of this approach is that the relative movements in the individual components of the series can be observed:

Weekly wage rates relative to 1950 (1950 = 100)

Type of worker	1950	1955	1960
Unskilled	100	122·9	165·7
Semi-skilled	100	128·2	164·1
Skilled	100	116·7	154·2
Clerical	100	122·5	130·0
	400	490·3	614·0

Using this method, his index of wage rates would be:

$$1950 \quad \frac{400}{4} = 100$$

$$1955 \quad \frac{490·3}{4} = 122·6$$

$$1960 \quad \frac{614·0}{4} = 153·5$$

Alternatively he may prefer to sum the wage rates themselves and then

divide by the number of rates before converting to an index numb
form (the 'aggregative' method). The sums are as follows:

1950, £40·50;   1955, £49·50;   1960, £62·00

Thus the indices will be:

$$1950 \quad \frac{£40·50}{4} \div \frac{£40·50}{4} \times 100 = \frac{£40·50}{£40·50} \times 100 = 100$$

(notice that both sums will always be divided by the same figure, whi
can be cancelled).

$$1955 \quad \frac{£49·50}{£40·50} \times 100 = 122·2$$

$$1960 \quad \frac{£62·00}{£40·50} \times 100 = 153·1$$

4.2.3   Both of these methods unfortunately omit one vital consideratio
The significance to the firm of these changes in the weekly wage rat
will depend upon the number of workers in each group. A large i
crease in a rate which is paid to only a handful of workers will ha
less influence on the total wages bill than a smaller increase for
category of workers involving hundreds of men. It seems that diff
ential importance must be given to each increase in wage rates c
pending upon the numbers of men affected. In the terminology
index numbers we shall *weight* the individual components.

To illustrate the application of weighting let us assume that the fi
has employed the same number of workers throughout the period a
that the average wages bill has been found as follows:

Type of worker	Number of workers (quantity weights)	Average wage bill (value weights)
Unskilled	9	£102·00
Semi-skilled	23	£293·25
Skilled	17	£252·17
Clerical	1	£11·75
	50	£659·17

4.2.4   Now we may apply these weights to the simple methods consider
in 4.2.2. We shall multiply each wage rate relative by the correspondi
value weight and each actual wage rate by the quantity weight and s
the products. Averaging will then produce the overall indices.

Type of worker	Average wage bill (value weight) (1)	Wage rates relative to			(1)×(2)	(1)×(3)	(1)×(4)
		1950 (2)	1955 (3)	1960 (4)			
Unskilled	£102·00	100	122·9	165·7	10200	12535·800	16901·400
Semi-skilled	£293·25	100	128·2	164·2	29325	37594·650	48151·650
Skilled	£252·17	100	116·7	154·2	25217	29428·239	38884·634
Clerical	£11·75	100	122·5	130·0	1175	1439·375	1527·500
	£659·17				65917	80998·064	105465·184

Just as the mean can be found from $\dfrac{\Sigma fx}{\Sigma f}$, so the average (weighted index) for the above situation is found by dividing the sum of products of indices and weights by the sum of the weights. This gives:

$$1950 \qquad \frac{65917}{659·17} = 100$$

$$1955 \qquad \frac{80998·064}{659·17} = 122·9$$

$$1960 \qquad \frac{105465·184}{659·17} = 160·0$$

Remembering, in the aggregative method, that the 50 will cancel out in the comparison of the current year and base year averages we find:

$$1950 \qquad \frac{£517·00}{£517·00} \times 100 = 100$$

$$1955 \qquad \frac{£634·50}{£517·00} \times 100 = 122·7$$

$$1960 \qquad \frac{£826·00}{£517·00} \times 100 = 159·8$$

4.2.5   A number of points emerge from these results. Firstly, we notice that the weighted average of the wage relatives does not produce the same answers as the indices based upon the weighted averages of the original wage rates. This is not surprising because of the fundamentally different approaches involved.

'Aggregative' method

Weekly wage rates

Type of worker	Number of workers (quantity weights) (1)	1950 (2)	1955 (3)	1960 (4)	(1)×(2)	(1)×(3)	(1)×(4)
Unskilled	9	£8·75	£10·75	£14·50	£78·75	£96·75	£130·50
Semi-skilled	23	£9·75	£12·50	£16·00	£224·25	£287·50	£368·00
Skilled	17	£12·00	£14·00	£18·50	£204·00	£238·00	£314·50
Clerical	1	£10·00	£12·25	£13·00	£10·00	£12·25	£13·00
	50				£517·00	£634·50	£826·00

It is easily appreciated if we consider in general terms the calculations undertaken. If we denote the wage rate (i.e. the price of labour) by p, the sets of weights by w and use the subscript $_B$ to indicate the base year and the subscript $_C$ to show the current year (the year for which the index is to be calculated) our calculations (ignoring throughout the factor of 100) are as follows:

	'Relatives' method	*'Aggregative'* method
Unweighted	$\dfrac{\sum \dfrac{p_C}{p_B}}{n}$	$\dfrac{\Sigma p_C}{\Sigma p_B}$
Weighted	$\dfrac{\sum \dfrac{p_C}{p_B} \times w}{\Sigma w}$	$\dfrac{\Sigma p_C w}{\Sigma p_B w}$

It can be seen that the two methods, both for the unweighted and weighted series, are not mathematically identical.

Secondly, we should notice that the weighted indices produce slightly higher answers (particularly for 1960) than the unweighted ones. This is the expected consequence of there being collectively more employees paid on the two rates subject to the greater increase than on the two which increased by a smaller amount.

4.3 Laspeyres and Paasche weighting

4.3.1 It was assumed in the wages example in 4.2.3 that the number of employees (used in the weights) remained unchanged throughout the period under review. The reader will appreciate that in many fields this is unlikely to be true. Technological progress, new methods, and changes in tastes may produce a proliferation of possible weights which are available to the statistician. Which weights should he use? We will again discuss the problem with reference to a simple example, involving this time the construction of a quantity (as opposed to a price) index series.

A car distributor requiring an impression of changes in the volume of his sales will necessarily want to take account not only of the numbers sold, but also the prices at which the sales were made. The sale of one limousine may be equivalent to four or five small family saloons. It seems that in this situation prices and total sales value provide weighting factors which will be applied to the quantity relatives actual numbers. Let us now consider the following data:

Model	April 1958 Numbers sold	Price (£hundreds)	April 1959 Numbers sold	Price (£hundreds)	April 1960 Numbers sold	Price (£hund)
A	30	4·5	45	4·5	42	5·0
B	5	6·5	5	6·8	6	7·0
C	1	20·0	2	20·5	3	21·0
	36		52		51	

With the series based upon April 1958, we require an index of volume of sales in April 1959 and April 1960. At first sight it may app that all the prices could be incorporated into the index. Using aggregative method for each year, the numbers of cars could be m plied by the respective prices and the total value of sales calculated 1959 and 1960 as a percentage of 1958. This would indeed be a q valid index number series, but it would relate to current value and no volume. If a volume index is to be found, then in each calculation only variable should be the number of cars sold. The weights should fixed and made to relate to both the base year and the current y quantities in the formula.

4.3.2   Many series regularly produced by official agencies use the base y data in formulating weights. This approach, first suggested by the C man economist Etienne Laspeyres, has a number of points to comm it. The weights, whether they be prices, quantities or anything else, h only to be established once; they can then be used during the rest of series. This is an important factor when one considers an index cover some aspect of the national economy: the execution of a survey obtain the weighting data would be both time-consuming and expens so it could only be undertaken occasionally. With a base weighted se the calculation of each index is also completed with greater speed, important consideration for series produced monthly, weekly or da

The major disadvantage to be set against these favourable point that the base weights may quickly become irrelevant in a dyna economy. The expiration of too long a period between the rebasing recalculation of weights for any series can quickly invalidate the resu an example being the *Interim index of retail prices* produced in

United Kingdom after 1947: this index series came in for heavy criticism because it used weighting based upon a survey of expenditure habits that was carried out in 1938. In price index systems it will be found (other things being equal) that as prices increase so quantities purchased will fall and vice versa, yet using base year quantities as weights this will not be reflected and the series will therefore tend to exaggerate the real magnitude of the price changes.

4.3.3   Let us now return to the car distributor example cited earlier and see how the base weighted series is obtained. We shall consider both the relatives and aggregative approaches.

'Relatives' method

Model	Base year value weights $q_{1958} \times p_{1958}$ (1)	Quantity relatives 1958 (2)	1959 (3)	1960 (4)	(1)×(2)	(1)×(3)	(1)×(4)
A	135·0	100	150	140	13500	20250	18900
B	32·5	100	100	120	3250	3250	3900
C	20·0	100	200	300	2000	4000	6000
	187·5				18750	27500	28800

The base weighted indices will therefore be:

$$\text{April 1958} \qquad \frac{18750}{187 \cdot 5} = 100$$

$$\text{April 1959} \qquad \frac{27500}{187 \cdot 5} = 146 \cdot 7$$

$$\text{April 1960} \qquad \frac{28800}{187 \cdot 5} = 153 \cdot 6$$

'Aggregative' method

Model	Base year price weights p_{1958} (1)	Number of cars sold 1958 (2)	1959 (3)	1960 (4)	(1)×(2)	(1)×(3)	(1)×(4)
A	4·5	30	45	42	135·0	202·5	189·0
B	6·5	5	5	6	32·5	32·5	39·0
C	20·0	1	2	3	20·0	40·0	60·0
					187·5	275·0	288·0

Using this approach we compare the value of sales in the current year at base year prices with the value of sales in the base year at base year prices. This gives:

April 1958 $\quad \dfrac{187 \cdot 5}{187 \cdot 5} \times 100 = 100$

April 1959 $\quad \dfrac{275}{187 \cdot 5} \times 100 = 146 \cdot 7$

April 1960 $\quad \dfrac{288}{187 \cdot 5} \times 100 = 153 \cdot 6$

4.3.4 It should be noticed that the two sets of results in the case of base weighting are identical. The reason for this can be seen from the formulae shown below.

Base weighting

	Quantity index	Price index
Relatives method	$\dfrac{\sum \frac{q_C}{q_B} \times q_B p_B}{\Sigma q_B p_B}$	$\dfrac{\sum \frac{p_C}{p_B} \times p_B q_B}{\Sigma p_B q_B}$
Aggregative method	$\dfrac{\Sigma q_C p_B}{\Sigma q_B p_B}$	$\dfrac{\Sigma p_C q_B}{\Sigma p_B q_B}$

With appropriate cancelling

$$\frac{\sum \frac{q_C}{q_B} \times q_B p_B}{\Sigma q_B \times p_B} \equiv \frac{\Sigma q_C p_B}{\Sigma q_B p_B}$$

4.3.5 It is now possible to indicate why different weighting systems are used with the relatives and aggregative methods. If simple price weights had been used with the relatives method an exaggerated effect would have been given to the increases in the sales of model C, resulting in indices of 171·8 for April 1959 and 239·0 for April 1960. The reason is that very high weight (20 out of a weighting total of 31) is being given to high quantity relatives (200 and 300 in 1959 and 1960) even though the percentage of revenue from model C was only increasing at 1958 prices from 11 per cent in 1958 to 15 per cent in 1959 to 21 per cent in 1960. It seems that if this potential exaggeration is to be reduced both prices and quantities must be incorporated in the weights to give the value figures used above.

4.3.6   The use of current weights which was suggested by H. Paasche remedies the criticisms directed at the Laspeyres system, because the weights used will be revised each year or month to take account of changes in tastes, output, etc. However, the collection of the necessary data to establish the current weights may prove impossible in practice, owing to the cost and time involved. Furthermore, the current series will tend to understate price changes at a time of rising prices, just as the base weighted series was shown to exaggerate them, whilst the calculation of each index using current weights will be very much more time-consuming than for a base weighted series in which the denominator of the final expression remains constant. The calculation of the current weighted indices for the quantity indices will demonstrate this:

'Relatives' method

Model	Current year value weights $(q_{1959} \times p_{1959})$ (1)	$(q_{1960} \times p_{1960})$ (2)	Quantity relatives 1959 (3)	1960 (4)	(1)×(3)	(2)×(4)
A	202·5	210·0	150	140	30375	29400
B	34·0	42·0	100	120	3400	5040
C	41·0	63·0	200	300	8200	18900
	277·5	315·0			41975	53340

The calculations for the base year are identical to those for the base weighted system and have therefore been omitted. The indices are as follows:

April 1959 $\dfrac{41975}{277 \cdot 5} = 151 \cdot 3$

April 1960 $\dfrac{53340}{315} = 169 \cdot 3$

Using the aggregative approach we compared the value of sales in the current year at current year prices with the value of sales in the base year at current year prices:

April 1958 $\dfrac{187 \cdot 5}{187 \cdot 5} \times 100 = 100$

April 1959 $\dfrac{277 \cdot 5}{189 \cdot 5} \times 100 = 146 \cdot 4$

April 1960 $\dfrac{315 \cdot 0}{206 \cdot 0} \times 100 = 152 \cdot 9$

'Aggregative' method

Model	Current year weights (prices) 1958 (1)	1959 (2)	1960 (3)	Numbers of cars sold 1958 (4)	1959 (5)	1960 (6)	(2)×(4)	(2)×(5)	(3)×(4)	(3)×(6)
A	4·5	4·5	5·0	30	45	42	135·0	202·5	150·0	210·0
B	6·5	6·8	7·0	5	5	6	34·0	34·0	35·0	42·0
C	20·0	20·5	21·0	1	2	3	20·5	41·0	21·0	63·0
							189·5	277·5	206·0	315·0

The general forms of the current weighted methods are as follows, indicating that the relatives and aggregative systems are not in this case the same.

Current weighting

	Quantity index	Price index
Relatives method	$\dfrac{\sum \dfrac{q_C}{q_B} \times q_C p_C}{\Sigma q_C p_C}$	$\dfrac{\sum \dfrac{p_C}{p_B} \times p_C q_C}{\Sigma p_C q_C}$
Aggregative method	$\dfrac{\Sigma q_C p_C}{\Sigma q_B p_C}$	$\dfrac{\Sigma p_C q_C}{\Sigma p_B q_C}$

4.3.7   Although the arithmetic mean has been used throughout this discussion so far, it is possible to use other averaging methods with index numbers. It was noted in 3.3.2 that in cases where relative changes are to be averaged then the geometric mean has distinct theoretical advantages. Thus the geometric mean of quantity relatives may be found from

$$\text{Index} = \Sigma w \sqrt{\left[\left(\frac{q_{C_1}}{q_{B_1}}\right)^{w_1} \times \left(\frac{q_{C_2}}{q_{B_2}}\right)^{w_2} \times \ldots \times \left(\frac{q_{C_n}}{q_{B_n}}\right)^{w_n} \right]}$$

or in log form

$$\log \text{Index} = \frac{\Sigma \log \left(\dfrac{q_C}{q_B}\right) \times w}{\Sigma w}$$

In practice the computational difficulties are deemed to outweigh the theoretical merits so that the majority of indices use the simpler arithmetic mean.

4.3.8   The weights used so far have been expressed in absolute form. Sometimes it is simpler to convert value weights in money terms to a relative version. We may find the weight for food in a price index given as 31·9 or 319 indicating that 31·9 per cent of total expenditure was on foodstuffs.

Finally, the reader should not conclude that only base or current year weights are used. Often neither method is applied, for there may be insufficient data to make either system feasible. A fixed weight series might then be constructed, perhaps using the results of a survey conducted neither in the base nor in the current year. Alternatively the

weights may be the average of several years' figures (possibly unrelat
to the base or current year). If these fixed weights are value weigh
they may be revalued at base year or current year prices/quantities.

4.4 The perfect index?

4.4.1 The imperfections of the Laspeyres and Paasche systems of ind
numbers have engaged the attention of many economists and stat
ticians. Several suggestions for a compromise solution have been p
forward, and two of these will now be considered. In essence, both i
volve the averaging of some component of the base and current ye
indices.

F. Y. Edgeworth's solution was to average the base and curre
weights to give for the price index:

$$\text{Index} = \frac{\Sigma p_C(q_B + q_C)}{\Sigma p_B(q_B + q_C)}$$

Rather more refined, and not a little more complicated, is the soluti
proposed by I. Fisher. His 'ideal index' rests upon the averaging of t
full Laspeyres and Paasche indices, using the geometric mean:

$$\text{Index} = \sqrt{\frac{\Sigma p_C q_B}{\Sigma p_B q_B} \times \frac{\Sigma p_C q_C}{\Sigma p_B q_C}}$$

4.4.2 For reasons of economy, it is unlikely that either of these two formul
will ever be employed in practice, but it is worth noting them becau
each satisfies some criterion for recognition as a perfect index numb
Few of the earlier methods which we have considered pass these tests f
perfection.

4.4.3 *The time reversal test*

4.4.3.1 The time reversal test stipulates that, with a perfect index numb
system, the reversal of the time subscripts should produce the reciproc
of the original index. Looked at in another way, the product of t
original index and its time amendment should equal unity. Consider t
following simple example involving fixed weights:

Items	Weights (quantities)	Prices 1946	1966	p_Bq_F	p_Cq_F
A	4	2	6	8	24
B	6	5	11	30	66
C	2	1	5	2	10
				40	100

Ignoring the multiplication by 100, the index for 1966 (1946 = 100) is given by:

$$\frac{\Sigma p_Cq_F}{\Sigma p_Bq_F} = \frac{100}{40} = 2 \cdot 5$$

Reversing the subscripts (in p_B and p_C) the index for 1946 (1966 = 100) is given by

$$\frac{40}{100} = 0 \cdot 4$$

$I_{1966(1946=100)} \times I_{1946(1966=100)} = 2 \cdot 5 \times 0 \cdot 4 = 1$. The test is therefore satisfied.

The aggregative index with fixed weighting is the only system considered earlier which will satisfy this test. It can quickly be shown that both the current weighted and base weighted series will fail:

$$\frac{\Sigma p_Cq_C}{\Sigma p_Bq_C} \times \frac{\Sigma p_Bq_B}{\Sigma p_Cq_B} \neq 1 \text{ (unless } q_C = q_B)$$

4.4.3.2 However, both Edgeworth's and Fisher's formulae satisfy this test. In the first case:

$$\frac{\Sigma p_C(q_B+q_C)}{\Sigma p_B(q_B+q_C)} \times \frac{\Sigma p_B(q_C+q_B)}{\Sigma p_C(q_C+q_B)} = 1$$

In the second case:

$$\sqrt{\frac{\Sigma p_Cq_B}{\Sigma p_Bq_B} \times \frac{\Sigma p_Cq_C}{\Sigma p_Bq_C}} \times \sqrt{\frac{\Sigma p_Bq_C}{\Sigma p_Cq_C} \times \frac{\Sigma p_Bq_B}{\Sigma p_Cq_B}} = 1$$

4.4.4 *The factor reversal test*

4.4.4.1 This test states that in a perfect index number system, the product of a price index (with quantity weights) and its quantity counterpart (with price weights) should produce an index of total value. We shall show that only Fisher's ideal index satisfies this test. For example, the Laspeyr aggregative system produces the following situation:

$$\text{Price index} = \frac{\Sigma p_C q_B}{\Sigma p_B q_B}; \quad \text{quantity index} = \frac{\Sigma q_C p_B}{\Sigma q_B p_B}$$

Therefore
$$\frac{\Sigma p_C q_B}{\Sigma p_B q_B} \times \frac{\Sigma q_C p_B}{\Sigma q_B p_B} = \frac{\Sigma p_C q_B \times \Sigma p_B q_C}{(\Sigma p_B q_B)^2}$$

which is not the same as the value index given by:

$$\frac{\Sigma p_C q_C}{\Sigma p_B q_B}$$

4.4.4.2 With Fisher's version:

$$\text{Price index} = \sqrt{\frac{\Sigma p_C q_B}{\Sigma p_B q_B} \times \frac{\Sigma p_C q_C}{\Sigma p_B q_C}}$$

$$\text{Quantity index} = \sqrt{\frac{\Sigma q_C p_B}{\Sigma q_B p_B} \times \frac{\Sigma q_C p_C}{\Sigma q_B p_C}}$$

The product of the two gives

$$\sqrt{\frac{(\Sigma p_C q_C)^2}{(\Sigma p_B q_B)^2}}$$

$$= \frac{\Sigma p_C q_C}{\Sigma p_B q_B} = \text{value index}$$

4.4.4.3 It does seem that mathematically the perfect index may exist. However, we should again stress that mathematical perfection and the economic and administrative realities of practice are unlikely to coincide when index numbers are involved. Expediency will usually win the day. As long as the computational side is reasonably straightforward, and providing the weighting reflects the approximate situation under study and the index itself is acceptably accurate, then the constructor of the series can feel quite happy that a good job has been done. The very nature and art of the economic statistician is compromise. He will rarely have completely adequate data, nor will he have the resources to collect additional data.

4.5 The base

4.5.1 We have not so far referred to the problems of deciding at which point in time the base of the series should be fixed, when it should be amended and what the implications of changing the base will be.

The base period which is eventually selected may be a particular day or the average for a month, a year or a number of years: all these alternatives have been utilized in index number construction. The main general criterion for fixing the base is that the variables under study should be reasonably stable, because only if this condition is satisfied will the indices for subsequent periods have real meaning. For instance, to set the base for a price index in a month when there was a general round of price increases will give the impression in later months of little or no increase in the cost of living. Similarly indices of wage rates may produce a misleading impression if the base of the series is set a month or two after a substantial pay award for the members of the nation's largest trade union organization. In practice it may be difficult to find a stable situation for all the variables to be included and once again a compromise solution has to be adopted. In any case, the creation of a completely new series may well determine the base period as the time when the first index can be compiled and published: under these circumstances the desire to produce results quickly may override the stability requirement.

4.5.2   Changing the base

4.5.2.1   While a series may continue for several years with a fixed base period, the time will eventually come when the base has to be brought up to date. The changeover, although mechanically straightforward, is complicated by the fact that long-run trends of figures are no longer immediately available. The reader requiring such a long run of data should not, however, be unduly alarmed. A simple means exists by which the two series may be linked. Suppose that we are faced with the following situation:

Index of meat prices

	(1 January 1952 = 100)
1 January 1955	110·3
1 January 1958	119·6
1 January 1961	132·1
1 January 1964	158·6
	(1 January 1964 = 100)
1 January 1965	102·3
1 January 1966	105·8

We have here the index for the old series for 1 January 1964 as well as the information that this date is the base of the revised series. We must be given these facts if the linking is to take place. Now we observe that meat prices in 1965 are 102·3 per cent of those in 1964. The old series figure for 1964 is 158·6 so that the figure for 1 January 1965 (1 January 1952 = 100) will be 102·3 per cent of 158·6, i.e. $\dfrac{102\cdot3}{100} \times 158\cdot6$

$= 162\cdot2$. Similarly the index for 1 January 1966 $= \dfrac{105\cdot8}{100} \times 158\cdot6 = 167\cdot8$

Alternatively the indices for the period 1952 to 1961 may be expressed in terms of 1 January 1964 = 100, in which case we ask what percentage of 158·6 each of these figures represents:

1 January 1952	$\dfrac{100}{158\cdot6} \times 100 = 63\cdot1$	
1 January 1955	$\dfrac{110\cdot3}{158\cdot6} \times 100 = 69\cdot5$	
1 January 1958	$\dfrac{119\cdot6}{158\cdot6} \times 100 = 75\cdot4$	
1 January 1961	$\dfrac{132\cdot1}{158\cdot6} \times 100 = 83\cdot3$	

4.5.2.2   We must now consider the possibility that these methods may be somewhat suspect. For instance, if the above example were derived from the aggregative method (with fixed quantity weights throughout the period) the procedure would have been as follows:

$$\frac{\Sigma p_{1965} q_F}{\Sigma p_{1964} q_F} \times \frac{\Sigma p_{1964} q_F}{\Sigma p_{1952} q_F} = \frac{\Sigma p_{1965} q_F}{\Sigma p_{1952} q_F}$$

This is correct. However, for a base or current weighted aggregative series or for any price relatives series, this simplification does not exist. (The reader may care to check this for himself.) In consequence one is not mathematically justified in linking the majority of published indices in this simple way. Once again, however, the loss in precision is often small and the advantages of a long run of figures considerable. The method, though imperfect, is widely employed.

4.5.4   *Chain-linked index numbers*

Although this chapter has, so far, considered the conventional methods of index number construction in which the variables in each year are expressed as a percentage of a fixed base, we should also notice that there will be circumstances when changes from month to month, or year to year, will prove of interest. The system of chain-linked indices achieves this by comparing a variable or group of variables in one month/year with the previous month/year.

Gross domestic product at 1958 constant prices (£ million)		Chained indices
1956	21,070	
1957	21,474	$\dfrac{21474}{21070} \times 100 = 101 \cdot 9$
1958	21,478	$\dfrac{21478}{21474} \times 100 = 100 \cdot 0$
1959	22,365	$\dfrac{22365}{21478} \times 100 = 104 \cdot 1$
1960	23,484	$\dfrac{23484}{22365} \times 100 = 105 \cdot 0$
1961	24,268	$\dfrac{24268}{23484} \times 100 = 103 \cdot 3$
1962	24,442	$\dfrac{24442}{24268} \times 100 = 100 \cdot 7$
1963	25,537	$\dfrac{25537}{24442} \times 100 = 104 \cdot 5$
1964	26,912	$\dfrac{26912}{25537} \times 100 = 105 \cdot 4$

Here we can see at a glance the picture of economic growth from year to year.

There are a number of advantages to be derived from this method, although it also has some limitations. Firstly, it is possible in a composite index to introduce new items from time to time, and to adjust the weights without the necessity of recalculating the whole series; also the individual calculations can be based upon any of the techniques which have been outlined earlier. One important restriction, however, is that only on the

assumption of fixed and unchanging weights and an aggregati
approach will the evaluation of changes from any period to any oth
period be possible. For instance, if a price index for 1960 is given by

$$\frac{\Sigma p_{1960} q_F}{\Sigma p_{1959} q_F}$$

and the index for 1961 by

$$\frac{\Sigma p_{1961} q_F}{\Sigma p_{1960} q_F}$$

then the product of the two will give the index for 1961 compared wi
1959, i.e.:

$$\frac{\Sigma p_{1960} q_F}{\Sigma p_{1959} q_F} \times \frac{\Sigma p_{1961} q_F}{\Sigma p_{1960} q_F} = \frac{\Sigma p_{1961} q_F}{\Sigma p_{1959} q_F}$$

If any other method is applied, as was the case with changing the bas
the necessary cancelling and simplification are not possible.

4.6 The United Kingdom index of retail prices

4.6.1 Although we have dealt with the main theoretical methods and looke
at some of the problems associated with the construction of inde
numbers, the examples used have for the sake of simplicity been largel
hypothetical. In order to fully realize the difficulties likely to be en
countered in the practical application of indices we must now examin
the development and compilation of regularly produced series. To thi
end, we shall first study the construction of a price index by examinin
the United Kingdom index of retail prices.

4.6.2   This index is designed to show month-to-month changes in the averag
prices of goods and services that are purchased by defined families
Although often referred to as a 'cost of living' index, this is an inaccurat
description as many goods are included which might be left out of a
series measuring living costs. Nevertheless, not every payment made i
included in the index, some of those omitted being:

(i) Income tax payments.
(ii) Purchase of savings certificates or payments to savings clubs.
(iii) National Insurance contributions, life insurance premiums an
superannuation payments.
(iv) Premiums on household insurance (other than for buildings).
(v) Subscriptions to trade unions, friendly societies, hospital funds
and other cash gifts.
(vi) Betting stakes.

(vii) Private medical fees.

(viii) Capital sums or mortgage payments for house purchase.

Their exclusion is justified on a number of grounds, the impossibility of identifying 'units' for which price changes can be recorded being the most important; in any case, many of the excluded payments constitute family savings or investment rather than family expenditure.

4.6.3   The index is a weighted arithmetic mean of price relatives. It uses percentage expenditure weights which are multiplied by ten (so that the sum of the weights is 1,000 rather than 100), and is based upon the following main groups of items:

    (i) Food.

    (ii) Alcoholic drink.

    (iii) Tobacco.

    (iv) Housing.

    (v) Fuel and light.

    (vi) Durable household goods.

    (vii) Clothing and footwear.

    (viii) Transport and vehicles.

    (ix) Miscellaneous goods.

    (x) Services.

Some variation in the number of groups employed has occurred over the years, but these are the ten groups currently in use.

In order to calculate the price relatives for these groups, each is divided into sections which number 91 in all. Within each section a representative list of goods and/or services has been built up and it is for each of these 350 items that prices are recorded. As an illustration of this breakdown, let us examine group (x), Services (from *Studies in Official Statistics*, no. 6, 'Methods of construction and calculation of the index of retail prices', Ministry of Labour and National Service [1959]) (see p. 150).

4.6.4   The prices of the 350 items which are recorded each month are those actually charged in cash transactions. In other words, no charges for hire purchase are counted, nor are discounts unless these are given to all customers. The only exception to the latter rule is the 'Co-op' dividend, which cannot be ascertained at the time of purchase and is therefore omitted. The collection of the prices ruling is undertaken either by personal visits to retailers by representatives of the Ministry of Labour or through postal contact with manufacturers, retailers and public corporations. The majority of the food items are covered by the first method, the data being collected by 200 Ministry of Labour offices situated in a

(x) Services

Section	Items
83. Postage.	(1) Inland letter post rate.
	(2) Poundage on postal and money orders.
84. Telephones, telegrams etc.	Rental for residential telephone; charges local, toll and trunk calls; call box charges.
85. Cinemas.	Charges for designated seats at evening p formances at cinemas in various towns.
86. Other entertainment.	(1) Admission to First Division football mat
	(2) Television set rental.
	(3) Radio and combined television-ra licences.
	(4) Admission to dance halls in various tow
	(5) Subscription to a youth club.
87. Domestic help.	Hourly rates in various towns.
88. Hairdressing.	(1) Men's haircut.
	(2) Women's shampoo and set.
89. Boot and shoe repairs.	(1) Men's shoes, soleing and heeling.
	(2) Women's shoes, heeling.
90. Laundry charges.	Bagwash; or charge for one sheet and o shirt.
91. Dry cleaning and miscellaneous services.	(1) Charge for cleaning man's suit.
	(2) Charge for cleaning woman's coat.
	(3) Charge for watch cleaning.

representative list of towns throughout the country. Prices are observ in the various forms of retail outlet such as sole traders, multiple r tailers, cooperatives, department stores etc., and when changes in eith the quantity or quality of the items observed take place an adjustme to the price is made, although this procedure is somewhat arbitrary.

4.6.5   Before discussing the methods of combining the various price chang in the 350 items, we must say something about the weighting facto which are used. This necessarily involves us in a discussion of t development of the series since 1945.

On the advice of the Cost of Living Advisory Committee establish by the Minister of Labour in 1947, a temporary series was started wi 17 June 1947 as the base period. This interim index sought to measu the changes in the prices of goods and services purchased by workin class families, and the weights were based upon a pre-war survey working-class family budgets. These were percentage expenditu weights, revalued at pricings ruling in June 1947. The weights related to only eight broad groups of items, drink and tobacco being combined in one group, whilst transport and vehicles were not separately specified. By 1951 the advisory committee had decided that expenditure habits had become sufficiently stable after the Second World War to justify a comprehensive survey of family expenditure in order to remedy the acknowledged defects in the weighting of the interim index. As there would be an inevitable delay in undertaking this survey and producing the results, a number of modifications to the interim index were made. The base period was shifted to 15 January 1952 and the weights revised to make them proportionate to the estimated consumption by working-class families in 1950, revalued at January 1952 prices. The groups were extended by separating alcoholic drink and tobacco.

4.6.6   The interim index of retail prices was replaced from 14 February 1956 by the full index of retail prices with a base period of 17 January 1956. The new index utilized the results of the household expenditure survey conducted during 1953–4 when 13,000 households kept detailed records of their expenditure. The weights were calculated from the returns of 11,638 households, the heads of which had a recorded gross weekly income of £20 or less, or which received more than 25 per cent of their total income from sources other than National Insurance pensions, National Assistance grants or other pensions. The expenditure patterns exhibited by these were adjusted where certain understatements were deemed to have occurred (in the light of other sources of data) and were revalued at January 1956 prices. The ten groups of items for weighting purposes were introduced at this time.

This series continued unaltered from February 1956 to January 1962. As with the interim index it was fundamentally base weighted (although fixed weighted is a more accurate description). From February 1962 to January 1963 the weights were revised, even though the base period was unaltered, and were related to the results (revalued at 1962 prices) of the annual family expenditure surveys conducted between 1958 and 1961. This pattern of the yearly revision of the weights in February on the basis of the last three family expenditure survey results has continued since this time, although from February 1963 the base period was revised to 16 January 1962 where it is expected to remain for a number of years. Another change affecting the calculation of weights is that the income limit for the head of the household has progressively been revised from £20 in the household expenditure survey to £30 and now £40 in the family expenditure survey. Although fixed weighting is still the correct term for the index of retail prices, the weighting structure now bears far closer affinity to a current weighted system than has previously been the case with United Kingdom government series.

4.6.7   The weights which have prevailed over the period covered by the two series are shown below:

Interim index of retail prices

	June 1947–January 1952	February 1952–January 1956
(i) Food	348	399
(ii) Rent and rates	88	72
(iii) Clothing	97	98
(iv) Fuel and light	65	66
(v) Household durable goods	71	62
(vi) Miscellaneous	35	44
(vii) Services	79	91
(viii) Drink and tobacco	217	78 (Alcoholic drink) 90 (Tobacco)

Index of retail prices

(*Each set of weights runs from February in the first year shown to January in the second*)

	1956–1962	1962–1963	1963–1964	1964–1965	1965–1966	1966–1967
(i) Food	350	319	319	314	311	298
(ii) Alcoholic drink	71	64	63	63	65	67
(iii) Tobacco	80	79	77	74	76	77
(iv) Housing	87	102	104	107	109	113
(v) Fuel and light	55	62	63	66	65	64
(vi) Durable household goods	66	64	64	62	59	57
(vii) Clothing and footwear	106	98	98	95	92	91
(viii) Transport and vehicles	68	92	93	100	105	116
(ix) Miscellaneous goods	59	64	63	63	63	61
(x) Services	58	56	56	56	55	56

These values are useful not only in the context of calculating this series but also because they throw considerable light on the changing pattern of relative expenditure over the post-war period (notice the steady reduction in the weight for food and the opposite movement in the weights for transport and vehicles and housing).

4.6.8   We can now consider the method of bringing together all the collected data to produce the group 'price relatives' and finally the all-items index. With few exceptions the several prices obtained for individual commodities and services are averaged (unweighted) and compared with the average price at the base period to produce an item price relative. These item indices have next to be grouped into their respective sections to produce 91 section price indices. Some weighting is employed at this stage because the item indices are used as price indicators to reflect the relative movements in other commodities. However, the weights are approximate and for some sections a simple average is preferred. Each group weight is now broken down into the various sections of the group as shown below for fuel and light.

	Weight 1956–1962
(v) Fuel and light	55
40. Coal	26
41. Coke	2
42. Gas	12
43. Electricity	12
44. Oil and other fuel and light	3

The usual approach is adopted in calculating the group indices, namely multiplying each section index by its weight, summing the products, and dividing by the sum of the weights. The price indices for the ten main groups are then similarly combined using the main-group weights to produce the all-items index.

4.6.9   The price data for this series is collected on one specific day. The Tuesday closest to the 15th of the month is selected unless this is the day after a bank holiday, in which case the previous or following Tuesday is chosen so that four or five weeks elapse between successive investigations. As it takes approximately five weeks to process and publish the results obtained on each index day the series is generally available in the third week of the following month, when a press release is issued. It then appears in the *Ministry of Labour Gazette*, the *Monthly Digest of Statistics* and the *Board of Trade Journal*. Complete coverage over a number of years is given for the ten main groups (together with the group weights) and the all-items index. In addition, the main groups are sometimes subdivided (not necessarily into the sections mentioned above) to give greater detail.

4.7 The United Kingdom index of industrial production

4.7.1 As an illustration of a quantity or volume index, we shall now consider the United Kingdom index of industrial production. This index, as prepared by the Central Statistical Office in collaboration with the statistics divisions of various ministries and government departments, is designed to measure the monthly changes in the volume of United Kingdom industrial production.

The definition of 'industrial' is in some ways rather restrictive, because mining, quarrying, manufacturing, construction, gas, electricity and water are included, but service industries (distribution, transport, finance and all other public and private services) plus agriculture, forestry and fishing are omitted. The exclusion of these items can be justified but nevertheless may account for apparent peculiarities in the value of indices at certain times. For instance, industrial production may appear stagnant at a time when unemployment is being reduced. The obvious implication is that the employment possibilities are being created in the service industries. However, for those industries included the output (from both private firms and state-controlled or government organizations) is recorded for the production of capital and consumption goods, whether they are for domestic consumption, for export, or for the armed forces.

In all 880 separate series are incorporated into this index. Each of these should show the difference between, on the one hand, the output of products and, on the other hand, the inputs of materials, products and services from other industries (plus an adjustment for changes in stocks of materials and part-manufactured goods). In practice, the necessary information is rarely available, so that output data (in the form of quantities or value), input data, or employment data have to be utilized on the assumption that changes in the amount of work completed will be proportionate to the changes in these variables. Although under conditions of instability one should have reservations about this assumption (particularly in the case of labour employed), in normal circumstances it would appear to be reasonably valid.

Fortunately, the majority of the series employed relate to the quantity of output delivered or produced (measured in tons, gallons, cubic yards etc.). Those data showing the value of output present a problem, in that these values have to be deflated by the use of appropriate price indices to the level of prices ruling at the base period; if this is not done an impression of value changes compared with the base period instead of volume changes will be given. It should be stressed, however, that the value series are rather more useful for industries producing a varied range of goods or where there is no suitable unit for measuring the output in physical terms. If in these circumstances value of output data is also unavailable, then figures of the inputs (i.e. consumption of raw materials and/or labour) will have to be used. For example, in the shipbuilding and construction industries, where the production of individual items extends over a long period, measurement of the work in progress is only possible on the basis of input data. The relative importance of the four types of series within the index is shown below:

Type of series	Percentage of series
Output	
Quantities delivered or produced	80
Value of deliveries or sales	12
Input	
Quantities of major materials received	7
Number of persons employed	1
	100

4.7.2   A further adjustment which has to be made to the original series is in connexion with the time interval to which they relate. The current situation is shown below:

Time interval	Percentage of series
Weekly	1
Weekly averages of four- and five-week periods	11·5
Calendar months	68
Quarters	8
Quarters (with alternative series for shorter intervals)	11·5

Index of industrial production*

Average 1958=100

| | Total all industries | Mining and quarrying | Manufacturing industries | | | | | | | | | | | | |
| | | | Total all manufacturing industries | Food, drink and tobacco | | | Chemicals and allied industries | | | Engineering and allied industries | | | | |
				Total	Food	Drink and tobacco	Total	Coke ovens, oil refineries etc.	General chemicals etc.	Total	Engineering and electrical goods	Shipbuilding and marine engineering	Vehicles (including aircraft)	Metal goods not elsewhere specified
Weights	*1,000*	*72*	*748*	*86*	*55*	*31*	*68*	*9*	*59*	*310*	*167*	*22*	*79*	*42*
1959	105·1	97	106·0	104	102	107	111	107	111	105	105	94	109	100
1960	112·5	94	114·6	107	104	112	123	117	123	113	114	85	118	112
1961	113·9	93	114·8	110	107	116	125	122	125	114	121	86	110	105
1962	115·1	95	115·3	112	110	116	129	120	130	115	123	87	112	101
1963	119·0	95	120·0	115	112	121	138	124	140	119	126	77	121	108
1964	128·2	95	129·5	118	112	129	151	135	153	128	137	75	126	121
1965	131·7	92	133·6	121	114	133	158	142	160	132	143	74	130	123
Unadjusted														
1964 October	135	100	137	124	119	134	156	137	158	133	142	75	132	127
November	137	103	139	128	120	141	158	146	159	136	148	76	129	132
December	130	95	129	118	110	132	148	139	149	134	151	76	123	114
1965 January	132	91	133	110	110	111	155	138	158	133	141	76	130	133
February	137	102	137	117	112	126	163	138	167	136	146	75	131	134

April	129	91	131	118	111	132	160	153	161	127	131	75	135	125
May	136	101	140	126	116	143	159	147	160	140	148	75	144	133
June	131	88	135	123	114	139	161	141	164	135	143	74	141	124
July	120	86	120	119	112	133	152	147	153	116	130	73	105	101
August	116	70	118	117	110	129	146	128	148	111	122	71	106	101
September	133	87	136	122	117	130	159	135	162	133	147	73	124	125
October	137	96	141	129	122	141	159	140	162	138	150	74	134	130
November	139	98	141	131	124	144	159	142	161	139	152	75	134	132
December	133	94	132	117	111	129	153	155	153	138	156	75	132	112
1966 January	133	79	133	112	111	113	159	145	161	134	145	74	129	128
Seasonally adjusted														
1964 October	129	94	130	120	113	131	152	138	155	128	137	75	128	120
November	130	95	132	121	114	134	154	144	155	130	140	75	129	123
December	132	97	133	121	113	135	154	138	157	133	145	75	130	121
1965 January	133	97	135	122	115	134	156	142	159	136	149	75	129	129
February	132	96	134	121	113	136	157	136	160	133	143	74	129	129
March	131	94	131	122	112	141	159	141	162	130	140	74	125	124
April	131	92	132	119	113	130	158	150	160	130	138	74	131	125
May	132	93	134	118	112	129	154	147	155	135	146	75	132	125
June	130	89	133	118	113	128	160	141	163	130	138	75	131	124
July	131	92	133	123	117	134	159	148	161	130	140	74	132	117
August	132	93	134	121	116	130	162	132	167	132	147	74	120	120
September	131	90	132	120	114	131	156	137	159	130	140	73	129	120
October	132	90	134	124	116	137	155	140	158	134	145	74	133	123
November	132	90	133	124	117	136	155	140	157	133	144	74	131	123
December	133	90	135	121	114	132	161	153	162	136	150	74	136	118
1966 January	133	89	135	124	116	137	160	149	162	138	153	73	130	125

* See footnote on page 159.

Index of industrial production* (continued)

Average 1958=100

	Metal manufacture			Textiles, leather and clothing				Bricks, pottery, glass etc.			Timber, furniture etc.	Paper, printing and publishing	Other manufacturing industries	Construction	Gas, electricity and water
								Manufacturing industries (continued)							
	Total	Ferrous	Non-ferrous	Total	Textiles	Leather, leather goods and fur	Clothing and footwear	Total	Bricks, cement etc.	Pottery and glass					
Weights	68	55	13	92	58	4	30	28	17	11	20	55	22	126	54
1959	104	104	107	107	106	103	111	106	107	106	112	107	108	106	103
1960	121	121	123	113	110	102	120	118	120	116	114	119	120	111	110
1961	114	113	116	111	107	102	122	123	129	115	116	120	116	120	116
1962	108	106	114	109	105	97	118	126	129	121	111	122	118	121	125
1963	113	111	119	112	110	99	120	130	131	127	113	128	126	121	133
1964	128	127	131	119	116	102	126	148	154	139	128	139	137	135	137
1965	134	134	133	121	118	103	129	150	158	139	127	142		138	145
Unadjusted															
1964 October	137	138	133	133	129	106	146	160	163	155	140	151	147	⎫	139
November	141	141	143	129	130	107	131	159	164	152	147	151	147	137	152
December	125	123	133	110	113	103	107	146	149	141	125	130	137	⎭	169
1965 January	141	141	137	117	118	107	118	150	156	140	121	150	135	⎫	175
February	145	144	147	128	124	108	138	152	160	141	126	146	145	137	175
March	137	137	137	124	121	108	133	154	159	146	130	141	141	⎭	164

April	[illegible]	[illegible]	[illegible]	[illegible]	[illegible]	[illegible]	[illegible]	[illegible]	[illegible]	[illegible]	[illegible]	[illegible]	[illegible]		[illegible]
May	143	144	141	124	123	108	127	159	172	138	129	154	148	138	127
June	134	133	140	116	115	105	121	150	162	132	140	138	145		115
July	115	115	115	109	106	90	119	148	157	133	123	114	133		111
August	115	116	109	106	104	81	114	133	145	113	107	135	125	138	110
September	135	135	137	131	124	110	147	155	164	141	148	140	151		129
October	139	140	135	137	130	111	154	159	163	152	132	155			140
November	141	141	138	129	127	110	135	157	159	153	130	154		138	175
December	122	119	132	114	114	101	115	140	142	136	122	135			178
1966 January	131	131	134	118	116	110	124	143	149	134	112	153			189
Seasonally adjusted															
1964 October	130	131	126	120	117	100	130	152	157	144	124	141	136		142
November	133	133	134	119	117	99	126	152	160	139	131	140	140	137	139
December	132	130	138	120	117	105	128	156	160	149	136	138	144		143
1965 January	138	138	136	121	118	103	129	154	162	142	134	144	137		143
February	138	138	140	121	119	102	125	152	160	139	132	140	141	139	148
March	131	130	133	116	116	102	119	152	157	144	135	135	139		148
April	135	136	133	118	117	102	123	153	160	141	128	138	141		142
May	136	136	137	122	120	104	130	150	162	130	126	144	141	137	139
June	133	132	136	120	118	104	128	148	155	135	134	140	140		136
July	134	134	131	122	120	98	128	147	158	130	122	143	146		138
August	135	137	129	124	121	104	134	151	160	138	118	142	147	137	139
September	132	132	130	121	118	105	129	151	157	141	134	142	139		147
October	132	133	128	123	118	105	137	150	156	141	118	144			143
November	132	133	129	119	114	103	131	149	155	141	116	143		138	159
December	129	127	137	124	119	103	139	149	152	143	133	144			150
1966 January	129	128	133	122	116	105	136	147	154	136	124	147			154

(Central Statistical Office)

* This index provides a general measure of monthly changes in total industrial production. The unadjusted index allows for variations in the length of calendar months so that it gives the average weekly rate of production in each month. The seasonally adjusted index allows for holidays and regularly recurring seasonal movements.

As some of the figures on which the index is based are available only with considerable delay, the index numbers for the last few months are always liable to be revised. Industries are grouped according to the *Standard Industrial Classification* 1958.

We have now stated that the index purports to show the monthly changes in production, so that short-interval series are preferred. However, even with calendar months the difficulty caused by differing numbers of working days from month to month has to be overcome. The differences are ironed out by the use of a standard month so that the index compares the average weekly rates of production in the different months. Although this important adjustment is made to the monthly (and quarterly) series, no adjustment is made to take account of the incidence of public holidays and the index may show a decline (or may not rise as fast as before) during December because of the loss of production on Christmas Day and Boxing Day.

4.7.3   Having discussed the series which are incorporated in the index of industrial production we can now look at the ways in which they are combined. For any month each production figure is expressed as a percentage of the average monthly production of this commodity in the base year (currently 1958). The 880 quantity relatives are then integrated using the Laspeyres base weighted method. The 880 individual weights (value weights as in the index of retail prices) are derived as follows. First, the net output of each industry defined at the census of production in 1958 is calculated as a proportion (per 1,000) of total net output of all participant industries at that date. Certain adjustments are made to the census definition of net output because it fails to exclude amounts paid for services rendered by industries outside the scope of the index, such as advertising and insurance. Next the industry weights are apportioned to the commodity series according to the most convenient method of estimating the relative importance of each within the industry. One problem is that of commodities which are produced outside the principal industry group. Theoretically a reallocation of weights between industries should be effected to take account of this factor, but in practice iron castings are the only commodity for which production outside the industry is considered important enough to warrant such an adjustment.

4.7.4   Once the quantity relatives have been obtained each month, the computation of the overall index, and indices for separate industries or orders of industries in the standard industrial classification, can be undertaken quickly to produce the type of analyses shown above (*Monthly Digest of Statistics*, March 1966).

The problem is that there are often considerable delays before the figures from firms and businesses all over the country are received. This can be seen from the following table which shows the percentage of data available at mid-January 1962 for preceding months (taken from

'The index of industrial production: change of base year to 1958', *Economic Trends*, no. 101, March 1962).

Information received at mid-January 1962 for:	Percentage of weight carried Total	Provisional	Final
1961 November	78	40	38
October	84	37	47
September	100	8	92
August	100	8	92
July	100	3	97

It is evident that if the index were to be published only when all the necessary data had been received and confirmed there would be up to a six-month delay in the computation and publication of each month's index. In practice, the index for a particular month is compiled about six or seven weeks after the end of the month so that the figure given (and those for earlier months) may need later revision.

4.7.5   In addition to the industrial production indices described above a monthly seasonally adjusted series is also compiled which takes account of public holidays and other causes of seasonal variations in production by the method described in chapter 6 (a refinement of the ratio method to a moving average trend). The seasonal components are under continuous scrutiny and incorporate all the latest available data.

4.7.6   Despite the considerable advances that have been made towards the perfection of basic data sources, the objectives and fundamental methods of the index of production have remained unaltered since its inception in 1948. The interim index, which commenced in 1948, was based on 1946. The net output of each industry, subject to the adjustment for services mentioned earlier (4.7.4), was taken from the 1935 census of production and projected to 1946 on the assumption that during this period of eleven years the net output had changed in the same proportion as the total wages bill. Since detailed changes in the wages bills were only known for broad groups of industries, the relative importance of subdivisions of industry were poorly evaluated.

Once the results of the first post-war census (taken in 1948) became available in 1952, the interim index was abandoned and the index of industrial production based on 1948 was substituted.

Since that time, as successive census results have been produced, revisions in the weighting and base year of the series have followed. In 1958 the base year was shifted to 1954 and in 1962 it was moved to its current year of 1958. These relatively frequent revisions are primarily motivated by a desire to maintain the accuracy of the weighting structure. In addition, it must also be appreciated that the revisions in the standard industrial classification in 1948 and 1958 have necessitated a reconstruction of the industry weights and indices; these in part have reduced the long-term nature of the series. Finally, it should be pointed out that once the results of the 1963 census of production become available yet another reweighting and rebasing can be expected.

4.8   Examples

Index numbers construction

(a) A tenants' association requires information about the relative increases in the rents paid to the housing department of a borough council since 1945. Data is collected from tenants and produces the following results:

Weekly averages

Type of housing	1945 Rent	Number of tenancies	1950 Rent	Number of tenancies	1960 Rent	Number of tenancies
A	50p	1280	70p	1280	120p	1280
B	75p	760	100p	750	165p	750
C	100p	410	125p	580	200p	590
D	150p	86	185p	400	250p	700
E	200p	70	245p	250	325p	950
F	400p	10	460p	190	500p	1000

Calculate aggregative indices for 1950 and 1960 with 1945 = 100, using firstly the Laspeyres method and then the Paasche method. Account for the differences in the answers obtained from these approaches.

(b) The following data taken from the *Monthly Digest of Statistics* (March 1966) relates to the United Kingdom index of retail prices:

Group	1963 weights	1963 monthly average group index (16 Jan. 1962 = 100)	1965 weights	1965 monthly average group index (16 Jan. 1962 = 100)
(i) Food	319	104·8	311	111·6
(ii) Alcoholic drink	63	102·3	65	117·1
(iii) Tobacco	77	100·0	76	118·0
(iv) Housing	104	108·4	109	120·5
(v) Fuel and light	63	106·0	65	114·5
(vi) Durable household goods	64	100·1	59	104·8
(vii) Clothing and footwear	98	103·5	92	107·0
(viii) Transport and vehicles	93	100·5	105	106·7
(ix) Miscellaneous goods	63	101·9	63	109·0
(x) Services	56	104·0	55	112·7

(i) Calculate the 'all-items' index for 1963 and 1965 with 16 January 1962 = 100. If the 'all-items' index for 16 January 1962 is 117·5, with 17 January 1956 = 100, find the indices for 1963 and 1965 based upon 17 January 1956.

(ii) Calculate the index for 'non-food items' for 1963 and 1965 with 17 January 1956 = 100. (Index for food = 110·7, with weight 350 for 16 January 1962.)

(c) Convert the following three series in order to achieve a basis for a direct comparison between the variables measured.

	Percentage change in weekly earnings compared with June 1956	Index of hours of work (Jan. 1956 = 100)	Index of output (Jan. 1956 = 100)	(June 195 = 100)
January 1956	−1·5	100	100	
June 1956	0	100·2	104·0	
January 1957	+0·8	98·6	102·5	
June 1957	+2·5	100·8	106·3	100
January 1958	+3·2	98·5		100·2

(d) Obtain index numbers of consumers' prices for (i) food, drink and tobacco as a combined group and (ii) housing, fuel and light as a combined group, for the years 1955, 1961 and 1963 with 1958 = 100 from the following data:

Consumers' expenditure

	Current prices (£ million)				Revalued at 1958 prices (£ million)			
	1955	1958	1961	1963	1955	1958	1961	1963
Food	4,045	4,547	4,955	5,293	4,367	4,547	4,838	4,929
Alcoholic drink	835	913	1,054	1,181	881	913	1,083	1,121
Tobacco	880	1,031	1,217	1,286	937	1,031	1,101	1,084
Housing	1,058	1,375	1,682	1,945	1,309	1,375	1,482	1,584
Fuel and light	527	686	784	991	637	686	751	898

(Taken from tables 18 and 19, *National Income and Expenditure* 1964)

Explain what form of index you have obtained. Compare your results with the price movements of all goods and services as shown below, and fully interpret your answer.

1958 = 100

1955	1958	1961	1963
90	100	104	109

(University of London)

(e) From the data shown below on the value and volume of retail sales in Great Britain, obtain an index of retail prices (all kinds of business) for the years 1962, 1963 and 1964 with 1961 = 100. Assuming the volume index is of current weighted form, state what form your price index is.

Average 1961 = 100

	1962	1963	1964
Index of sales value	103	108	113
Index of sales volume	100	103	106

(University of London; table 112, *Monthly Digest of Statistics*, March 1966)

Chapter 5
Two-variable linear regression and correlation

Everyday experience constantly demonstrates the manner in which apparently different phenomena are associated or related in some way. Familiarity with these relationships enables us to plan ahead and predict the likely outcome of situations and events which have yet to occur. In this way we are able to use the evidence of the past in making decisions about the future with greater certainty and confidence than guesswork alone provides.

5.1 The practical application of relationships

5.1.1 Consider the obvious connexion between the size of a house and the cost of upkeep. The appreciation of this relationship will usually deter the prospective buyer of a large property who has the necessary capital but only a modest income. Equally apparent is the relationship between the speed of a car and the distance travelled. Does not a driver, recognizing this fact, add a few miles an hour to his speed when trying to catch his morning train after oversleeping? Finally, let us look at the problem of the housewife taking decisions about what and how much to buy. Generally, we will find that perishable foodstuffs are purchased in limited quantities only. The more rapid the deterioration the smaller will be the quantity purchased and the more often will individual purchases be made: the relationship implicit here is recognized by the housewife when she compiles her shopping list.

5.1.2 These examples could, of course, be extended indefinitely, but enough has been said to indicate the considerable use which everyone makes of relationships between variables and the logic of extending this idea to the specific problems faced by economists, sociologists or industrialists will be apparent. For instance, can it be shown that investment and employment are related, that there is a connexion between socio-economic status and juvenile delinquency, or that sales and advertising expenditure move together? If the answer is 'yes', then projections based on these relationships can be used in evaluating the implications of change. Wasted time, effort and finance will have been avoided.

5.1.3   In this chapter we shall, for two reasons, consider only linear relationships, or situations where a unit change in one variable will result in a constant absolute change in the other. In the first place the analysis of associated series in a linear form is relatively straightforward, yet nevertheless demonstrates the fundamental principles which are applicable to more complex situations where the relationship is better expressed by some other form of mathematical curve (see *Statistics for the Social Scientist: 2 Applied Statistics*). Secondly a linear relationship between two variables is a surprisingly common occurrence, and even where a refined non-linear curve might prove slightly superior the simpler form will often be quite adequate in the context of the problem under consideration.

5.2   The scatter diagram

5.2.1   It will be rare in the field of behavioural or social sciences to find two series of figures which are related perfectly by a straight line: it is more likely that only a general linear pattern or tendency will be apparent. The reasons for this are not difficult to find.

Suppose one is studying the relationship between the research and development expenditure and the profit margin on products of a number of firms. While it may be generally true to state that the two will increase together, it is inevitable that some firms' profit margins will be higher than others with the same R and D expenditure, and vice versa. The conditions under which the various firms are operating may be very different. The goods being produced, the firm's share of the market, the efficiency of the firm etc. will all play a part in determining the individual results.

To take another example, there may be a definite link between the age of workers in a factory and the number of days per year each man is absent from work because of illness. It will not be a precise relationship, however, as workers of the same age will have differing environmental factors affecting their general level of health. Housing, dietary habits, contact with illness and hereditary considerations will cause variations from person to person, and consequently variations from a perfect linear relationship.

5.2.2   We have indicated in 5.2.1 one of the principal differences between the social and the physical sciences. In an experiment in the physics or chemistry laboratory, more or less constant external conditions can be maintained, so that any relationships existing will be observed in approximately the pure form. In the social sciences, the very opposite applies. Here external conditions will never be constant and, as in the examples cited, there will always be exogenous factors influencing the relationship under review.

5.2.3   The type of situation confronting the statistician in the non-physical sciences is that shown by the bivariate data (observations having two simultaneous sets of data) below. It relates to the evaporation loss from fifty-gallon drums of aviation spirit which have been held in stock for varying periods of time. Ten drums, randomly selected, were examined:

Drum	Length of time in stock (days)	Evaporation loss (pints)
1	5	25
2	27	63
3	8	31
4	2	8
5	16	50
6	15	35
7	12	19
8	7	20
9	21	64
10	3	22

The first step in ascertaining the relationship between the time in stock and the evaporation loss is to plot a *scatter diagram*. This is a simple graphical device involving the location of one variable, designated y, on the vertical axis and the other, designated x, on the horizontal axis. Traditional mathematical terminology calls the dependent variable y and the independent variable x. While these definitions are generally adequate for our purposes, we should note the danger of inferring a causal relationship between y and x. We may say that numerically y depends on x, but it does not necessarily mean that y is caused by x. We shall elaborate on this point later in the chapter, but in the meantime it may be better to look upon y as the variable to be estimated on the basis of a given value of x.

In our case, it is common sense to realize that we are more likely to be interested in the effect of time on evaporation loss than the other way round. Thus evaporation loss is the y variable and time the x variable.

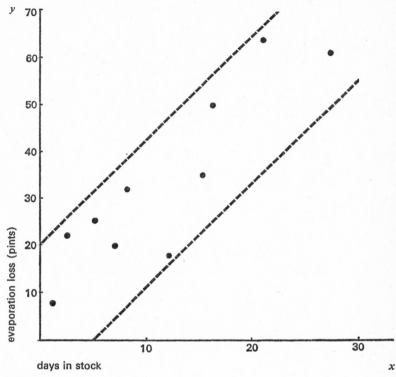

Fig. 50   The linear 'pattern'.

The ten pairs of values resulting from the examination of the drums can now be plotted on the scatter diagram as shown in fig. 50. As expected, an overall linear pattern emerges and the problem is to superimpose on this pattern the general relationship between y and x in the linear form, i.e. $y = a+bx$, which will remove the effect of outside factors.

5.2.4   Before developing this line of thought any further, let us recapitulate on the fundamentals of the linear equation as set out in chapter 1. In fig. 51 we have a straight line $y = a+bx$. The constant a is the value of y when x is equal to zero; in other words, it is the numerical distance between the origin and the point of intersection of the y axis and the line. The constant b is the value of the tangent to the angle formed between the line and any horizontal drawn from it.

In fig. 51 $b = \tan \theta = \dfrac{MN}{NL}$

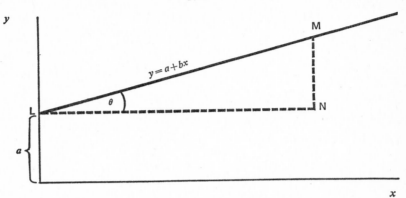

Fig. 51

5.2.5   It is evident that if we can draw in a line which adequately fits the data in our problem the estimation of the constants will not be a difficult matter. The real difficulty lies in finding the *line of best fit*. Even with the greatest care and skill a line drawn between the points with a ruler will be highly subjective and different individuals will arrive at different conclusions about the position. Nevertheless, in cases where the plotted points on the scatter diagram are more or less in a straight line, or where the precision of the estimates is not critical, this method can be used.

5.3   The method of least squares

5.3.1   Using the 'free-hand' method of curve fitting to represent the relationship portrayed on the scatter diagram, one tends, consciously or subconsciously, to draw in the straight line such that there are equal numbers of points located on either side. What is more the eye will automatically try to judge and equate the total distances between the points above and below the line. This is a conditioned recognition of the fact that the line of best fit must be an 'average' line in the true sense. Just as the sum of the deviations round the arithmetic mean of a series must equal zero, so the positive and negative deviations round a line of best fit (or regression line) must cancel out. This is the first of the conditions or requirements for an optimal line.

Unfortunately, it is not the only one. There are an infinitely large number of straight lines which will satisfy this condition. Three of these are shown on fig. 52. In all three cases the sum of the vertical deviations is zero as shown below (there is in fact a slight positive or negative sum in each one due to rounding in the equations).

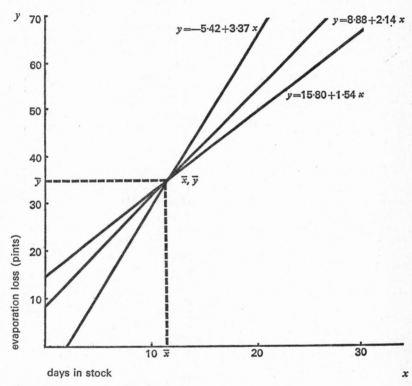

Fig. 52   Three equations where $\Sigma (y-y')=0$.

5.3.2   Two things emerge from this. Firstly, it appears that we need some second criterion for establishing a unique position for the regression line. Secondly, we find that the three lines considered above intersect at a common point. This is where lines drawn from the arithmetic mean of the x's (i.e. 11·6 days) and the arithmetic mean of the y's (i.e. 33·7 pints) cut each other. We will simply mention at this time that for any of the possible regression lines to be 'average' lines, they must inevitably pass through this point $\bar{x}, \bar{y}$.

171   Two-variable linear regression and correlation

Evaporation loss Actual y	Calculated y'	Deviations (y − y')
For y = 8·88+2·14x		
25	19·58	+5·42
63	66·66	−3·66
31	26·00	+5·00
8	13·16	−5·16
50	43·12	+6·88
35	40·98	−5·98
19	34·56	−15·56
20	23·86	−3·86
64	53·82	+10·18
22	15·30	+6·70
		0
For y = 15·80+1·54x		
25	23·52	+1·48
63	57·46	+5·54
31	28·14	+2·86
8	18·89	−10·89
50	40·49	+9·51
35	38·95	−3·95
19	34·32	−15·32
20	26·60	−6·60
64	48·20	+15·80
22	20·43	+1·57
		0
For y = −5·42+3·37x		
25	11·44	+13·56
63	85·62	−22·62
31	21·56	+9·44
8	1·32	+6·68
50	48·53	+1·47
35	45·16	−10·16
19	35·04	−16·04
20	18·18	+1·82
64	65·39	−1·39
22	4·70	+17·30
		0

5.3.3   The reader may recognize the similarity between the problem outlined here and that of finding a satisfactory method of measuring the dispersion of a series round its arithmetic mean. The solution to both is fundamentally the same. In calculating the standard deviation, the positive and negative deviations round the mean are squared, leaving positive answers in all cases. We do the same thing with the vertical deviations of the plotted points round our possible regression lines.

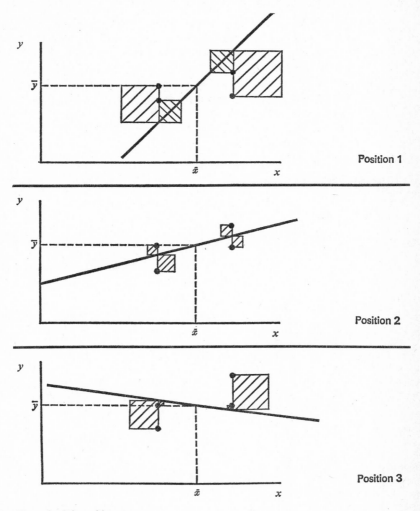

Fig. 53   The principles of least squares.

Our interest in this case, however, is not simply in achieving a non-zero sum: it is the magnitude of the sum which is our main concern. In fig. 53 we show diagrammatically the differences in the sum of squared deviations for different equations of a straight line. In position 1 the shaded areas are relatively large, but as the line is pivoted (on the $\bar{x}$, $\bar{y}$ point) in a clockwise direction to positions 2 and 3 the areas first become smaller and then begin to increase once more. There would seem to be some unique position at which the sum of the square deviations is at a minimum. This is the position of *least squares*. If we can ascertain the location of this particular straight line in terms of the constants *a* and *b* in the linear equation, then we have found the line of best fit. The reasoning is straightforward enough. Using the standard-deviation analogy once more, we know that the smaller the sum of the squared deviations round the mean the less dispersed is the series. In the context of regression analysis, we require the scatter or dispersion of points to be as small as possible. It follows that this will be achieved when the sum of squares is minimized.

5.3.4 *The normal equations*

5.3.4.1 Let us now consolidate our conclusions. Denoting by *d* the vertical deviations of the bivariate observations round the regression line, we must satisfy the following criteria, in order to discover the line of best fit:

$$\Sigma d = 0$$
$$\Sigma d^2 = \text{minimum}$$

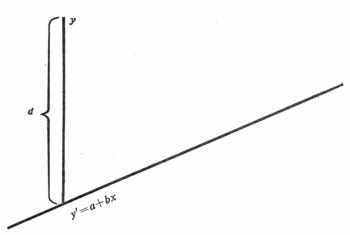

Fig. 54

But the deviation d is equal to the difference between the y value of the plotted point and the y' value on the regression line for the same value of x (this is shown in fig. 54):

$$d = y-(a+bx)$$

The second criterion can now be stated as the need to minimize

$$\Sigma\{y-(a+bx)\}^2$$

which requires the use of the differential calculus (see 5.8.1). By partially differentiating with respect of a (b being constant) and of b (a being constant) and simplifying, the following two normal equations are found which must be solved simultaneously for a and b:

(i) $\Sigma y = na+b\Sigma x$
(ii) $\Sigma xy = a\Sigma x+b\Sigma x^2$

5.3.4.2   It is evident that we need to calculate Σx, Σy, Σx^2 and Σxy, which can then be substituted in the above equations. This has been done below for the example cited in 5.2.3.

x	y	x^2	xy
5	25	25	125
27	63	729	1,701
8	31	64	248
2	8	4	16
16	50	256	800
15	35	225	525
12	19	144	228
7	20	49	140
21	64	441	1,344
3	22	9	66
116	337	1,946	5,193

Substituting these totals we get:

$$337 = 10a+116b \qquad \text{(i)}$$
$$5193 = 116a+1946b \qquad \text{(ii)}$$

Multiplying equation (i) by 11·6 and subtracting it from (ii) gives:

$$5193 = 116a+1946b$$
$$3909{\cdot}2 = 116a+1345{\cdot}6b$$
$$\overline{1283{\cdot}8 = \quad 0 + 600{\cdot}4b}$$

Therefore
$$b = \frac{1283 \cdot 8}{600 \cdot 4} = 2 \cdot 14$$

Using this value of b in equation (i) above leaves

$$337 \doteq 10a + (2 \cdot 14)(116)$$

So that $\quad a = \dfrac{337 - (2 \cdot 14)(116)}{10} = 8 \cdot 88$

Thus $y = 8 \cdot 88 + 2 \cdot 14x$, which is one of the equations graphed in fig. 52.

5.3.6 *The formula method*

For those readers who prefer not to have to solve the simultaneous equations we can derive formulae for finding a and b.

The course taken in solving the normal equations for a and b was to multiply (i) by $\dfrac{\Sigma x}{n} = \bar{x}$ (in our case 11·6). It should be noted that the mean of the x's will always be the appropriate multiplier. The result algebraically was as follows:

$$\frac{\Sigma x}{n}\Sigma y = \frac{\Sigma x}{n}na + \frac{\Sigma x}{n}b\Sigma x$$

Simplifying this and subtracting from equation (ii) gave

$$\Sigma xy = a\Sigma x + b\Sigma x^2$$

$$\frac{\Sigma x \Sigma y}{n} = a\Sigma x + \frac{b(\Sigma x)^2}{n}$$

$$\overline{\Sigma xy - \frac{\Sigma x \Sigma y}{n} = 0 + b\Sigma x^2 - \frac{b(\Sigma x)^2}{n}}$$

Re-arrangement of this expression leaves (see 5.8.2)

$$b = \frac{n\Sigma xy - \Sigma x \Sigma y}{n\Sigma x^2 - (\Sigma x)^2}$$

In our example $\quad b = \dfrac{(10)(5193) - (116)(337)}{(10)(1946) - (116)^2} = 2 \cdot 14$

Having calculated b we can, as before, substitute the result into the first of the normal equations which will leave a expressed in terms of the known numerical values:

From $\Sigma y = na + b\Sigma x$

we get $\quad a = \dfrac{\Sigma y - b\Sigma x}{n}$

so that $\quad a = \dfrac{337 - (2 \cdot 14)(116)}{10} = 8 \cdot 88$

The regression equation is therefore identical:

$$y = 8 \cdot 88 + 2 \cdot 14x$$

5.3.7 Simplified calculations

5.3.7.1 In many examples which the reader may encounter, the amount of arithmetic involved in the fundamental calculations of Σx, Σy, Σx^2 and Σxy is extensive and without the aid of a desk calculator rather tedious. For this reason we will now consider a way of simplifying this aspect of the regression problem. Basically the principle invoked is the same as that which applies to the calculation of the arithmetic mean using the assumed mean method.

If we subtract a constant from each value of x and each value of y (not necessarily the same constant), this will in no way affect the slope of the regression line but it will expedite the arithmetic. Let us illustrate this point by looking at a simplified example relating to the number of pupils per teacher in twelve secondary schools and the percentage of

Number of pupils per teacher x	Percentage of pupils passing three G.C.E. subjects y
17	5
19	6
12	8
10	7
16	5
13	7
18	3
20	4
22	5
13	9
19	3
8	11

177 **Two-variable linear regression and correlation**

students gaining at least three G.C.E. 'O' level subjects. Using the methods outlined in 5.3.6 the regression equation has been calculated and superimposed on the scatter diagram shown in fig. 55.

Now if we subtract 15 from each value of x and 6 from each value of y the only effect is to shift the origin of the axes. In relation to the regression line, the plotted points are not changed in any way. Therefore the amended figures can be utilized in the same way as the originals in calculating the slope of the line.

$u = (x-15)$	$v = (y-6)$	u^2	uv
+2	−1	4	−2
+4	0	16	0
−3	+2	9	−6
−5	+1	25	−5
+1	−1	1	−1
−2	+1	4	−2
+3	−3	9	−9
+5	−2	25	−10
+7	−1	49	−7
−2	+3	4	−6
+4	−3	16	−12
−7	+5	49	−35
+7	+1	211	−95

Thus $b = \dfrac{n\Sigma uv - (\Sigma u)(\Sigma v)}{n\Sigma u^2 - (\Sigma u)^2}$

giving $b = \dfrac{(12)(-95)-(7)(1)}{(12)(211)-(7)^2} = \dfrac{-1147}{2483} = -0.46$

It is evident that the amount of arithmetic has been considerably reduced. Unfortunately if we try to do the same thing in calculating a we will be making a fundamental mistake. According to the earlier formula we might try to find a from

$$a = \frac{\Sigma v - b\Sigma u}{n}$$

which gives $a = \dfrac{1-(-0.46)(7)}{12} = +0.35$

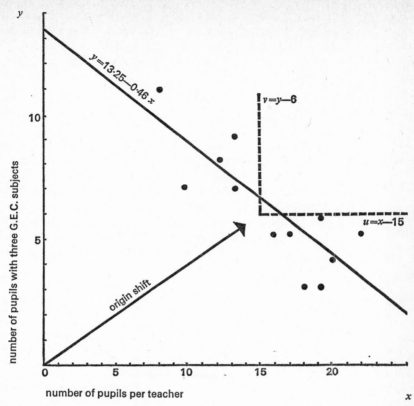

Fig. 55   The change in origin.

This is correct for the relationship between u and v, as is shown in fig. 55, but the problem is that we want the regression equation expressed in terms of the original units of measurement, x and y. Various methods can be used to effect this conversion but perhaps the best is to utilize the previously mentioned fact that the regression line passes through the point $\bar{x}$, $\bar{y}$. This can easily be demonstrated. If both sides of the first of the normal equations given in 5.3.5 are divided by n we get

$$\frac{\Sigma y}{n} = \frac{na}{n} + b\frac{\Sigma x}{n}$$

which reduces to $\bar{y} = a + b\bar{x}$

When x is equal to $\bar{x}$ then y must be equal to $\bar{y}$. If we find these two means and substitute them in the above equation we can establish the true value of a (note that $\bar{x}$ and $\bar{y}$ can be found as follows: $\bar{x} = 15 + \frac{7}{12}$ $= 15 \cdot 583$, $\bar{y} = 6 + \frac{1}{12} = 6 \cdot 083$). Now $6 \cdot 083 = a + (-0 \cdot 46)(15 \cdot 583)$ so that $a = 6 \cdot 083 - (0 \cdot 46)(15 \cdot 583) = 13 \cdot 25$ as before. The regression equation is

$$y = 13 \cdot 25 - 0 \cdot 46x$$

5.3.7.2   Let us now critically interpret the answers arrived at in the two examples considered earlier in this chapter (5.2.3, 5.3.7.1).

In the regression equation $y = 8 \cdot 88 + 2 \cdot 14x$, which expresses the relationship between evaporation loss from drums of aviation spirit and the length of time these drums have been stored, the value of b indicates that an increase of one unit in the length of time in stock will be accompanied by a 2·14-unit increase in the loss from evaporation. The figure for a suggests, on the other hand, that the drums were just over a gallon below capacity on arrival for storage. Although this may be correct, we should also consider the possibility that extrapolation is producing misleading results.

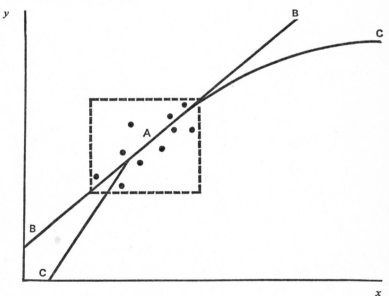

A = region of interpolation
B = regions of extrapolation
C = true relationships in regions of extrapolation

Fig. 56   The extrapolation trap.

Extrapolation, or the making of estimates or predictions outside the range of the experimental data, can be very unwise in certain circumstances. While a set of observations may show a good linear relationship between the variables, there is never any guarantee that the same linear form is present over those ranges of the variable not under consideration. It may be that a different linear equation fits the lower ranges or that a non-linear form is a better representation over the higher ranges. If this is believed to be the case, the only solution is to increase the number of observations in an attempt to fill in the blank areas on the scatter diagram. This argument is illustrated in fig. 56. It will be seen that the greatest care needs to be exercised in making statements based on extrapolation, but an awareness of the limitations of the technique should enable the user to avoid this particular trap.

Our conclusions about the relationship between 'O' level passes and the student/teacher ratio, expressed by $y = 13 \cdot 25 - 0 \cdot 46x$, differ from those above in only one respect. It was found that b had a negative value. This simply shows that the regression line has a negative slope; that is, the variables in question are moving in opposite directions. A one-unit increase in the number of pupils per teacher in these circumstances will result in a $0 \cdot 46$-unit decrease in the percentage of successful G.C.E. candidates.

5.4 Regressing y on x and x on y

5.4.1 In statistical literature the reader will often encounter the term 'the regression of y on x'. It may be that a certain equation expresses the regression of food expenditure on income, or total cost on output. This is no more than a form of shorthand showing that the first variable mentioned (food expenditure or total cost) is being estimated in terms of the second (income or output). In other words, if only the variables are mentioned in this way then for computational purposes the first will always be designated y and the second x.

5.4.2 It may appear that we are being unnecessarily pedantic in stressing this point. Surely, we may ask, once the line of best fit has been established either y or x can be estimated. Unfortunately this is not the case. If we are confronted with a situation where we both require to predict y from x and x from y, then two distinct equations need to be found. The regressions of y on x and x on y for the same bivariate data are not identical.

We can best demonstrate this by considering the following example.

A survey of eight families in different income groups is undertaken and the following data relating to expenditure habits collected.

| | Weekly per capita expenditure (pence) | |
Family	Protein	Carbo-hydrates
A	70·00	38·75
B	92·50	27·50
C	13·75	30·00
D	120·00	31·25
E	19·00	20·00
F	107·50	41·25
G	50·00	26·25
H	31·25	28·75

Now if protein expenditure is to be estimated, this is the y variable and carbohydrate expenditure is the x variable. Therefore

$$\Sigma y = 494{\cdot}00, \ \Sigma x = 243{\cdot}75, \ \Sigma x^2 = 7751{\cdot}56 \ \text{and} \ \Sigma xy = 16169{\cdot}06$$

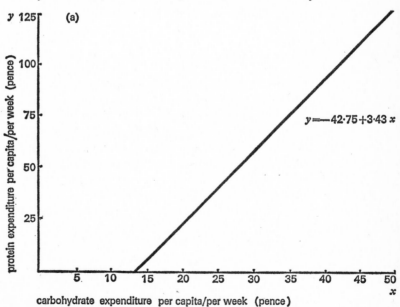

Fig. 57(a) Protein expenditure on carbohydrate and vice versa.

Using the formula method,

$$b = \frac{(8)(16169 \cdot 06) - (243 \cdot 75)(494 \cdot 00)}{(8)(7751 \cdot 56) - (243 \cdot 75)^2} = 3 \cdot 43$$

$$a = \frac{494 \cdot 00 - (3 \cdot 43)(243 \cdot 75)}{8} = -42 \cdot 75$$

The equation is $y = -42 \cdot 75 + 3 \cdot 43x$. This has been plotted in fig. 57(a).

If the object had been to estimate carbohydrate expenditure, however, the variable symbols would be reversed so that:

$$\Sigma y = 243 \cdot 75, \Sigma x = 494 \cdot 00, \Sigma x^2 = 4168 \cdot 91, \Sigma xy = 16169 \cdot 06$$

giving

$$b = \frac{(8)(16169 \cdot 06) - (494 \cdot 00)(243 \cdot 75)}{(8)(4168 \cdot 91) - (494 \cdot 00)^2} = 0 \cdot 10$$

$$a = \frac{243 \cdot 75 - (0 \cdot 10)(494 \cdot 00)}{8} = 24 \cdot 30$$

Here the equation is $y = 24 \cdot 30 + 0 \cdot 10x$, which is shown in fig. 57(b).

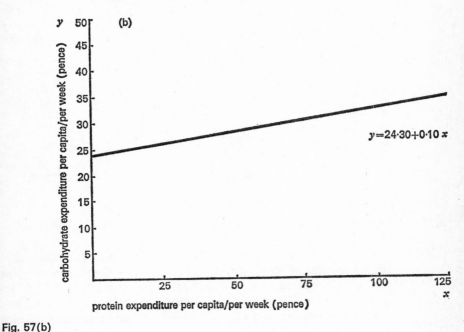

Fig. 57(b)

5.4.3   Although it is strictly correct to show the two regression lines on separate graphs it is often more convenient to consider them together.

Evidently we cannot designate both variables x and y simultaneously, so we must arbitrarily fix one as y and the other as x. If per capita expenditure on protein is measured on the vertical axis (y) and per capita expenditure on carbohydrates on the horizontal axis (x), we must modify the second of the regression equations. Although $y = -42.75 + 3.43x$ still represents the first relationship, the second must be changed to $x = 24.30 + 0.10y$. The two are shown together in fig. 58. It is immediately apparent that there is quite a considerable difference between them which, if ignored, would lead to erroneous estimates being made.

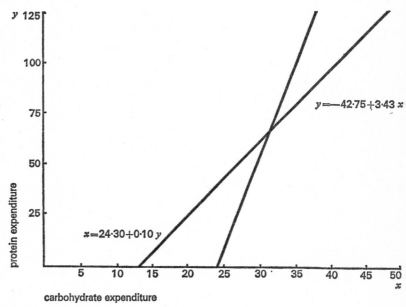

Fig. 58   The regression of y on x and x on y.

5.4.4   The difference between the two regression lines results from the underlying principles of the least-squares technique. We have seen that the sum of the squared vertical deviations must be minimized when regressing y on x. When on the other hand the regression of x on y is calculated, it is the sum of the squared horizontal deviations which must be as small as possible. Only in the event of the plotted points lying in a perfect straight line will the two equations be the same and the regression lines coincide.

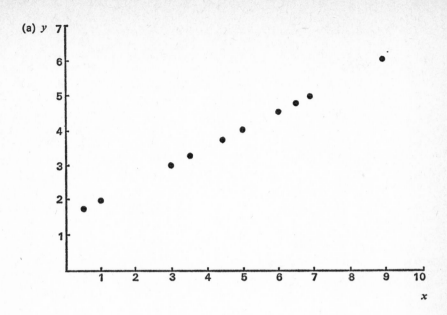

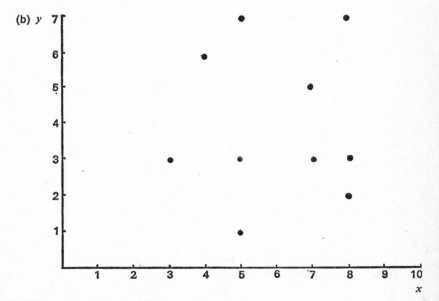

Fig. 59   The extremes of association.

185   Two-variable linear regression and correlation

5.5 The coefficient of correlation

5.5.1 Even though a calculated regression line is mathematically the best fit of the data under review, it provides no information about the strength of the relationship between the two variables. Fig. 59 shows two scatter diagrams. In the first case (a) there is a perfect linear relationship present while in the second (b) the absence of any apparent relationship is striking. The bivariate data seem to be completely random. In consequence the calculation of a regression equation for (b) will have no value or application at all. If, however, it was to be calculated and decisions were based upon the estimates of y, there would be a high probability of the wrong decisions being made.

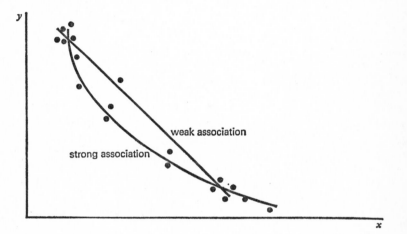

Fig. 60   Non-linear association.

5.5.2 It follows then that before the statistician goes to the trouble of establishing the least-squares line, he will require some means of evaluating whether numerically there is some connexion between the phenomena being studied. The *coefficient of correlation* provides this measure of the degree of linear association present. We should stress the word 'linear' once again in this connexion. It may be that there exists a strong non-linear relationship between the variables, and the calculation of the coefficient of linear correlation will, not unnaturally, fail to show this. In addition it may give to the unwary the impression that no form of association exists at all. This can well be a mistaken impression, as can be seen from fig. 60.

5.5.3   How are we to measure the degree of linear association? Let us look at fig. 61, in which we show a set of bivariate data and the mean lines, $\bar{y}$ and $\bar{x}$.

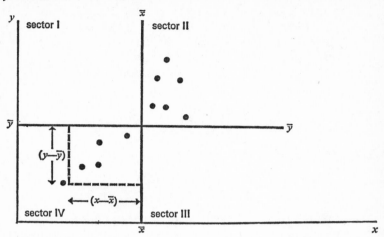

Fig. 61

If we multiply together the deviations of the x and y values round their respective means, i.e. $(x-\bar{x})(y-\bar{y})$, we will get a positive answer for values in sectors II and IV and negative answers for values in sectors I and III. If all these individual answers are summed then:

(1) When all or most of the points are in sectors II and IV, as in the illustration, a high positive answer results.

(2) When all or most of the points are in sectors I and III we obtain a high negative answer.

(3) When some of the points are in I and III and some are in II and IV then the positive and negative items more or less cancel out, leaving a small positive or negative sum.

It can be seen that the situation described under heading (1) above is that in which there is a strong association present and the slope of the regression line is positive. Heading (2) also shows the presence of a strong association, but here y decreases as x increases. Finally heading (3) suggests the existence of a random distribution of points on the scatter diagram, meaning that there is little or no association between y and x.

5.5.4   We have used the words high and low to describe $\Sigma(x-\bar{x})(y-\bar{y})$ (the expression $\dfrac{\Sigma(x-\bar{x})(y-\bar{y})}{n}$ is described as the covariance of x, y). The

187   Two-variable linear regression and correlation

problem is that the actual total will vary according to the number of observations and the specific location of the data with respect to the means. What we need is a method of reducing the absolute answer into a relative form so that comparisons can be made from situation to situation. To achieve this we divide $(x-\bar{x})$ by the standard deviation of the x distribution and $(y-\bar{y})$ by the standard deviation of the y distribution. This takes care of the first difficulty; the second is solved by averaging. The end result is the relative measure called the *Pearsonian* or *product moment coefficient of correlation*.

$$r = \frac{\Sigma (x-\bar{x})(y-\bar{y})}{ns_x s_y}$$

This in its turn reduces to the following (see also 5.8.3):

$$r = \frac{n\Sigma xy - \Sigma x \Sigma y}{\sqrt{[\{n\Sigma x^2 - (\Sigma x)^2\}\{n\Sigma y^2 - (\Sigma y)^2\}]}}$$

The reader will quickly understand why this last formula is preferred. The totals Σx, Σy, Σx^2 and Σxy are needed in ascertaining the regression line of y on x. The only additional calculation involved is therefore Σy^2. The other thing which should be noticed at this stage is that the calculated coefficient prescribes the strength of the association regardless of whether the regression equation to be established is that of y on x or x on y. In the formula shown above the reversal of the variable symbols will leave the structure completely unaltered.

5.5.5   Have we, in fact, reduced our answers to a relative form? We will now try to show that we have. It will be found that the coefficient will always lie within the range zero to one. A zero coefficient will cause us to infer the absence of any association between the variables. Conversely a figure of one represents the perfect relationship where the points on the scatter diagram lie exactly in a straight line. Most coefficients in practice will be between these two limiting cases. The interpretation of these intermediate values will be considered later in the chapter (5.7.1).

Some coefficients will be preceded by a positive and some by a negative sign. The reasons for and implications of this have already been covered, but it is nevertheless worth repeating that the sign of the coefficient shows only the type of relationship between y and x and gives no information about the degree of association. It is a common mistake among the uninitiated to interpret a value of -0.95 as being a situation of very weak association. The reader will reject this inference at all costs.

Let us now return to the calculation of the coefficients for the data shown in fig. 59.

In fig. 59(a)

y	x	y^2	x^2	xy
1·75	0·5	3·0625	0·25	0·875
2·00	1·0	4·0000	1·00	2·000
3·00	3·0	9·0000	9·00	9·000
3·25	3·5	10·5625	12·25	11·375
3·75	4·5	14·0625	20·25	16·875
4·00	5·0	16·0000	25·00	20·000
4·50	6·0	20·2500	36·00	27·000
4·75	6·5	22·5625	42·25	30·875
5·00	7·0	25·0000	49·00	35·000
6·00	9·0	36·0000	81·00	54·000
38·00	46·0	160·5000	276·00	207·000

$$r = \frac{(10)(207)-(38)(46)}{\sqrt{[\{(10)(276)-(46)^2\}\,\{(10)(160\cdot5)-(38)^2\}]}}$$

$$= \frac{322}{\sqrt{\{(644)(161)\}}} = \frac{322}{322} = +1\cdot0$$

This is the one limiting case, and a situation somewhat unlikely to be encountered in practice. Similarly, for the data shown in fig. 59(b) which demonstrates the other extreme:

y	x	y^2	x^2	xy
3·0	3·0	9·0	9·0	9·0
6·0	4·0	36·0	16·0	24·0
7·0	5·0	49·0	25·0	35·0
3·0	5·0	9·0	25·0	15·0
1·0	5·0	1·0	25·0	5·0
5·0	7·0	25·0	49·0	35·0
3·0	7·0	9·0	49·0	21·0
7·0	8·0	49·0	64·0	56·0
3·0	8·0	9·0	64·0	24·0
2·0	8·0	4·0	64·0	16·0
40·0	60·0	200·0	390·0	240·0

$$r = \frac{(10)(240)-(40)(60)}{\sqrt{[\{(10)(390)-(60)^2\}\,\{(10)(200)-(40)^2\}]}} = \frac{0}{\sqrt{\{(300)(400)\}}} = 0$$

We have thus outlined the method of calculation and demonstrated that the coefficient's limiting values are zero and one. Although usually our calculated answers will lie within these limits, we have at least shown the theoretical existence of $r = 0$ and $r = 1$. Also we have pointed out the stages involved in the computational procedure.

5.5.6 *Simplified calculations*

5.5.6.1   In 5.3.7 we saw that the calculations involved in finding the regression coefficient, b, could be simplified by subtracting arbitrary constants from the x and y values. The same approach applies to the calculation of the coefficient of correlation. In this case, however, the simplification can be taken one stage further.

In addition to changing the origin, it is also permissible in the context of correlation to change the unit of measurement. This means that where appropriate the u and v values can be further reduced by the use of an arbitrary divisor or multiplier, which need not be the same for both variables. This is demonstrated below in the calculation of r, between the diameter of bronze shafts produced on a lathe and the production time taken by the machinist. The figures result from observations taken at times randomly selected throughout the day.

Average diameter (inches) y	Average production time (seconds) x	$y - 3{\cdot}010$	$x - 200$	$v = 1000(y - 3{\cdot}010)$	$u = (x - 200)/5$	v^2	u^2	uv
3·006	195	− 0·004	− 5	− 4	− 1	16	1	+4
3·012	205	+0·002	+5	+2	+1	4	1	+2
3·001	200	− 0·009	0	− 9	0	81	0	0
2·998	185	− 0·012	− 15	− 12	− 3	144	9	+36
3·015	210	+0·005	+10	+5	+2	25	4	+10
3·009	215	− 0·001	+15	− 1	+3	1	9	− 3
3·013	200	+0·003	0	+3	0	9	0	0
3·000	190	− 0·010	− 10	− 10	− 2	100	4	+20
2·997	195	− 0·013	− 5	− 13	− 1	169	1	+13
3·005	205	− 0·005	+5	− 5	+1	25	1	− 5
3·010	200	0·000	0	0	0	0	0	0
3·016	220	+0·006	+20	+6	+4	36	16	+24
				− 38	+4	610	46	+101

Now $$r = \frac{n\Sigma uv - \Sigma u . \Sigma v}{\sqrt{[\{n\Sigma u^2 - (\Sigma u)^2\} \{n\Sigma v^2 - (\Sigma v)^2\}]}}$$

so that in this example

$$r = \frac{(12)(101)-(-38)(4)}{\sqrt{[\{(12)(46)-(4)^2\}\{(12)(610)-(-38)^2\}]}} = \frac{1364}{\sqrt{3149536}} = +0 \cdot 769$$

The arithmetic is very much simpler at all stages than it would have been if the original data had been used without amendment. The final calculations in that case would have been:

$$r = \frac{(12)(7277 \cdot 105)-(36 \cdot 082)(2420)}{\sqrt{[\{(12)(489150)-(2420)^2\}\{(12)(108 \cdot 493050)-(36 \cdot 082)^2\}]}} = +0 \cdot 769$$

5.5.6.2   Why can we use this second device in the correlation problem when we ignored it as a possibility in dealing with the regression coefficient? The answer can be found from a study of the formulae. Unless the unit of measurement of both x and y is changed by the same constant, the proportionate relationship between the numerator and denominator of

$$b = \frac{n\Sigma xy - \Sigma x \Sigma y}{n\Sigma x^2 - (\Sigma x)^2}$$

will be altered. The slope of the regression line is changed and will be incorrect. This is shown schematically in fig. 62.

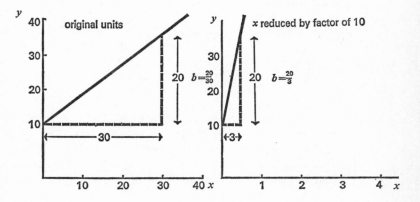

Fig. 62

The only remedy for this is to correct the answer by multiplying the value of b (found from the formula) by the amendment to the y's over the amendment to the x's. In the context of fig. 62 the answer of $\frac{20}{3}$ would be multiplied by $\frac{1}{10}$ giving $\frac{2}{3}$, which is correct.

191   Two-variable linear regression and correlation

Now the correlation coefficient is constructed in part by dividing $(x-\bar{x})$ and $(y-\bar{y})$ by their respective standard deviations. But it can be shown that if the values of a series are reduced by a specific amount then the standard deviation of the series is similarly reduced, e.g.:

x	$(x-\bar{x})^2$
50	0
20	900
80	900

$$\text{S.D.} = \sqrt{\frac{1800}{3}} = 24\cdot495$$

Reducing the x's by 10 gives

x	$(x-\bar{x})^2$
5	0
2	9
8	9

$$\text{S.D.} = \sqrt{\frac{18}{3}} = 2\cdot4495$$

Thus in

$$r = \frac{\Sigma(x-\bar{x})(y-\bar{y})}{ns_x s_y}$$

both the numerator and denominator will alter proportionately if the unit of measurement of x and y is changed.

It seems that the regression coefficient is the absolute ratio of the effect on y occasioned by a unit change in x. The coefficient of correlation, however, is a relative measure expressing the degree of linear association between y and x.

5.6 Grouped data

5.6.1 All the examples which have been cited so far have contained approximately 8–12 observations. Where these small groups of bivariate data are in evidence the techniques of calculation which have been outlined are perfectly adequate. If, however, 20 or more paired values of the two variables are available it is not difficult for the reader to imagine just how tedious the mechanical side of the computation will become.

Thinking back to chapter 2 of the book, we faced the same difficulty with the ungrouped data of a single series. The solution in that case, which we will also adopt here, was to construct a frequency distribution. In the present circumstances, the requirement will be for a 'double' frequency distribution, since we must group the two variables simultaneously while maintaining the pairs of values intact. This can be done by setting out the class intervals of one series vertically and the other horizontally.

5.6.2   The table of data below showing the age and earnings of the staff in a large firm demonstrates this form of presentation.

		y age (years)				column				
Income						1	2	3	4	5
(£ p.a.) x	20–30	30–40	40–50	50–60	60–70	$v = \dfrac{y - 1125}{250}$	f	fv	fv^2	fvu
2000–1750					3_{18}	3	3	9	27	18
1750–1500				3_4	2_8	2	5	10	20	14
1500–1250				2_2	1_3	1	3	3	3	4
1250–1000			6_0	7_0		0	13	0	0	0
1000–750	1_2	8_8	4_0	$2-_3$		−1	15	−15	15	8
750–500	2_8	3_6				−2	5	−10	20	14
500–250	5_{30}	1_3				−3	6	−18	54	33
row 1 $u = \dfrac{x-45}{10}$	−2	−1	0	1	2	Totals	50	−21	139	91
2 f	8	12	10	14	6	50				
3 fu	−16	−12	0	14	12	−2				
4 fu^2	32	12	0	14	24	82				
5 fuv	40	17	0	6	28	91				

It will be seen that from the ungrouped data we have found 5 men who are between 20 and 30 years old with earnings between £250 and £500, 2 men between 20 and 30 earning £500 to £750 and so on. Notice that the vertical distribution, giving the class intervals of earnings, has an unconventional descending order of magnitude from top to bottom, so that the overall form is that of a graph with the frequency figures providing what we may think of as a 'numerical scatter diagram'. We can therefore obtain at a glance some impression of the existence of the linear relationship.

5.6.3   The procedure for working out the regression equation and the co-efficient of correlation is as follows:

1. Establish the midpoints of the class intervals, and subtract the middle one from all the others.

2. Divide the deviations round this 'assumed mean' by a convenient figure (usually this will be the numerical width of the class interval if all the class intervals are equal).

3. Write down the end product of these manipulations in the column (1) and row (1) marked v and u respectively.

4. Sum the figures in the rows and the figures in the columns and enter the answers appropriately in column (2) and row (2). These are the frequencies of the two separate distributions.

5. Calculate Σfu, Σfu^2 and Σfv, Σfv^2 in the usual way (columns 3 and 4 and rows 3 and 4).

6. To find Σfvu, and as a check Σfuv, multiply each individual cell frequency by the value of v in the same row and by the value of u in the same column. Thus for the bottom left-hand corner fvu will be

$$5 \times -3 \times -2 = 30.$$

Enter each result in the corner of the cell. When all the fvu's have been calculated add up the numbers in each row and column (for the bottom row it will be $30+3 = 33$), place the answers in column (5) and row (5) and sum. Check that $\Sigma fvu = \Sigma fuv$.

5.6.4   Now instead of Σx, Σy, Σx^2, Σy^2 and Σxy we have Σfu, Σfv, Σfu^2, Σfv^2 and Σfuv. For purposes of substitution into the relevant formulae these are equivalent so that

$$r = \frac{(50)(91)-(-2)(-21)}{\sqrt{[\{(50)(82)-(2^2)\}\{(50)(139)-(-21^2)\}]}}$$

$$= \frac{4508}{\sqrt{26660864}}$$

$$= +0\cdot87$$

$$b\text{ (uncorrected)} = \frac{(50)(91)-(-21)(-2)}{(50)(82)-(2^2)} = \frac{4508}{4096} = 1\cdot1$$

As mentioned in 5.5.6.2 we must correct this answer (the regression coefficient) for the changes made in the units of measurement of both y and x (250 and 10 respectively). Following the rule already established, the uncorrected answer will be multiplied by 250/10 giving $b = 27\cdot5$ as the correct value.

Now $a = \bar{y}-b\bar{x}$, and $\bar{y} = 1125+\left(250 \times -\dfrac{21}{50}\right) = £1,020$,

also $\bar{x} = 45+\left(10 \times -\dfrac{2}{50}\right) = 44\cdot6$ years,

so that $a = 1020-(27\cdot5)(44\cdot6) = -206\cdot5$
The regression of y on x is
$$y = -206\cdot5+27\cdot5x.$$

5.7 The interpretation of r

5.7.1 In passing, we have already mentioned some of the potential traps into which it is easy to fall when looking at the question of the association between two variables. Let us summarize these and also extend the list.

(i) A low coefficient of linear correlation does not necessarily mean that there is no association at all. It may be that some non-linear regression line fits the data with a high degree of correlation. Alternatively, a low value of the coefficient may have resulted from the exclusion of a second or third independent variable which is important. A more elaborate model would be needed in this case.

(ii) A high coefficient does not necessarily mean that there is in reality strong association present or that the association is meaningful. From six throws of two dice one may get the following results:

Throw		i	ii	iii	iv	v	vi
Dice	I	6	2	4	1	5	3
	II	6	2	4	1	5	3

Obviously they have occurred by chance. Further throws would soon show the absence of any relationship and the presence of a purely random situation. A nonsense or spurious correlation is another phenomenon which must be watched for under this heading. We may find a high coefficient from data relating to the arrival of migrating storks at a village and the number of births there. Again, the number of divorces in Britain may be highly correlated with the sales of frozen and canned foodstuffs. In both it is apparent that the association is purely accidental: there is absolutely no connexion between the two variables. This reinforces our earlier argument that cause and effect should only be inferred when additional background evidence of a specialized nature is available.

5.7.2 Total, explained and unexplained variation

5.7.2.1 If it is decided to interpret r in terms of cause and effect we really need to know just how much of a change in the values of y is accounted for (explained) by the regression of y on x. If we can do this we will get a better impression of the strength of the linear association. It is common sense in thinking about the total variation in y to relate it to $\bar{y}$, so let us define the total variation in y as being $\Sigma(y-\bar{y})^2$. Of this total variation, $\Sigma(y'-\bar{y})^2$ is explained by the regression equation while $\Sigma(y-y')^2$ is purely random and cannot be explained in any systematic manner; it has resulted from the presence of factors or influences in the situation which are not related to, and cannot be explained by knowing, x. It is a residual independent of x.

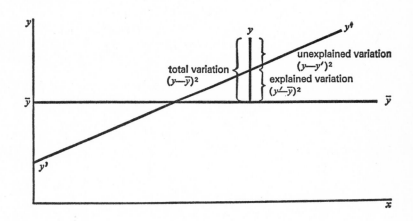

Fig. 63

Now if the explained variation is nearly as large as the total variation, then we have an indication that a change in y is closely linked to a change in x. If the reverse is true our conclusion will be that y is largely independent of x. The fraction

$$\frac{\text{explained variation}}{\text{total variation}} = \frac{\Sigma(y'-\bar{y})^2}{\Sigma(y-\bar{y})^2}$$

seems to provide a measure of causal association between y and x and is called the *coefficient of determination*.

5.7.2.2   If we work out the coefficient of determination in the example below, an interesting fact will emerge. The table shows the average number of cars per hour passing a check point on an arterial road out of a large city and the mean temperature for the same day. The observations were taken on ten Sundays during the summer months.

Average number of cars passing check point per hour (thousands)	Mean temperature (°F)
18	56
20	62
9	49
21	65
25	74
13	51
14	50
31	82
24	59
26	67

The regression of number of cars on temperature is given by

$$y' = -15 \cdot 20 + 0 \cdot 574x$$

The mean of the y distribution is $20 \cdot 1$. From these two facts we find the following:

x	y'	$(y' - \bar{y})^2$	y	$(y - \bar{y})^2$
56	16·944	9·960336	18	4·41
62	20·388	0·082944	20	0·01
49	12·926	51·466276	9	123·21
65	22·110	4·040100	21	0·81
74	27·276	51·494976	25	24·01
51	14·074	36·312676	13	50·41
50	13·500	43·560000	14	37·21
82	31·868	138·485824	31	118·81
59	18·666	2·056356	24	15·21
67	23·258	9·972964	26	34·81
		347·432452		408·90

The coefficient of determination is

$$\frac{347 \cdot 43}{408 \cdot 90} = 0 \cdot 85$$

Expressing this in a percentage form we can say that 85 per cent of the variance in y is attributable to the regression of y on x. In other words 15 per cent of the variability of y is caused by factors other than x (factors not taken into account in the two variable regressions). Now the point of interest mentioned earlier is that the coefficient of correlation for this data is $+0.922$, which if squared gives 0.85.

Now, it is a provable fact (see 5.8.4) that

Coefficient of determination $=$ (Coefficient of correlation)^2

The rather time-consuming process of calculating the coefficient of determination as set out in 5.7.3 can thus be avoided. In any case, the calculations could be radically simplified to provide an alternative method of calculating the coefficient of linear correlation by using the fact that:

$$\text{explained variation} = \Sigma(y'-\bar{y})^2 = a\Sigma y + b\Sigma xy - \frac{(\Sigma y)^2}{n} \text{ (see 5.8.5)}$$

$$\text{and} \quad \text{total variation} = \Sigma(y-\bar{y})^2 = \Sigma y^2 - \frac{(\Sigma y)^2}{n}$$

In this context these derivations have much to commend them. We require no more tabulated calculations than are involved in the estimation of a, b and r (using the product-moment approach), and the method possesses far greater generality than the earlier development, which is restricted to the two-variable linear model.

5.7.2.3 Some interesting conclusions can now be drawn, and our definitions of high and low correlation must be modified. Only if $r \geqslant 0.7$ is 50 per cent more of the variance of y caused by the regression of y on x, so a coefficient of correlation value below this can hardly be thought of as high, when outside factors are affecting y more than x does. For the real association to be as high as 95 per cent, r must equal 0.97 to 0.98; when the correlation coefficient is less than 0.5 then the association is really very low.

5.8 **Mathematical notes**

5.8.1 *Minimizing the sum of squares*

Section 5.3.4 indicated that to minimize the sum of the squares of the vertical deviations round the regression line, the differential calculus was required. The method is as follows:

To minimize $\Sigma[y-(a+bx)]^2$ the partial derivatives of the sum with respect to a and b should both be zero. This gives:

minimize $\qquad\qquad \Sigma(y-a-bx)^2$

Now $\quad \dfrac{\delta}{\delta a}\Sigma(y-a-bx)^2 = -2\Sigma(y-a-bx) = 0 \qquad$ (i)

and $\quad \dfrac{\delta}{\delta b}\Sigma(y-a-bx)^2 = -2\Sigma x(y-a-bx) = 0 \qquad$ (ii)

(i) simplifies to $\qquad\qquad 2\Sigma y = 2na + 2b\Sigma x$

$\qquad\qquad\qquad\qquad\qquad \Sigma y = na + b\Sigma x$

(ii) simplifies to $\qquad\qquad 2\Sigma xy = 2a\Sigma x + 2b\Sigma x^2$

$\qquad\qquad\qquad\qquad\qquad \Sigma xy = a\Sigma x + b\Sigma x^2$

These are the normal equations used earlier in the chapter.

5.8.2 *The formulae for b and a in the regression equation*

Given the normal equations

$$\Sigma y = na + b\Sigma x \qquad \text{(i)}$$
$$\Sigma xy = a\Sigma x + b\Sigma x^2 \qquad \text{(ii)}$$

multiplication of (i) by $\dfrac{\Sigma x}{n}$ and subtraction from (ii) leaves

$$\Sigma xy = a\Sigma x + b\Sigma x^2$$

$$\frac{\Sigma x \Sigma y}{n} = a\Sigma x + \frac{b(\Sigma x)^2}{n}$$

$$\Sigma xy - \frac{\Sigma x \Sigma y}{n} = b\Sigma x^2 - \frac{b(\Sigma x)^2}{n}$$

$$\frac{n\Sigma xy - \Sigma x \Sigma y}{n} = \frac{nb\Sigma x^2 - b(\Sigma x)^2}{n}$$

Multiplying both sides by n we obtain

$$n\Sigma xy - \Sigma x \Sigma y = nb\Sigma x^2 - b(\Sigma x)^2$$
$$n\Sigma xy - \Sigma x \Sigma y = b[n\Sigma x^2 - (\Sigma x)^2]$$

so that $\quad \dfrac{n\Sigma xy - \Sigma x \Sigma y}{n\Sigma x^2 - (\Sigma x)^2} = b$

Finally rearrangement of (i) given that b has been found enables a to be calculated:

$$a = \frac{\Sigma y - b\Sigma x}{n}$$

5.8.3 *The computational formula for r*

From the basic form

$$r = \frac{\Sigma(x-\bar{x})(y-\bar{y})}{ns_x s_y}$$

expansion gives:

$$r = \frac{\Sigma(xy+\bar{x}\bar{y}-x\bar{y}-\bar{x}y)}{n\sqrt{\left[\left\{\dfrac{\Sigma x^2}{n}-\left(\dfrac{\Sigma x}{n}\right)^2\right\}\left\{\dfrac{\Sigma y^2}{n}-\left(\dfrac{\Sigma y}{n}\right)^2\right\}\right]}}$$

$$= \frac{\Sigma xy+\dfrac{n\Sigma x\Sigma y}{n^2}-\dfrac{\Sigma x\Sigma y}{n}-\dfrac{\Sigma y\Sigma x}{n}}{\sqrt{\left[\left\{n\Sigma x^2-(\Sigma x)^2\right\}\left\{\dfrac{\Sigma y^2}{n}-\left(\dfrac{\Sigma y}{n}\right)^2\right\}\right]}}$$

$$= \frac{\Sigma xy-\dfrac{\Sigma x\Sigma y}{n}}{\sqrt{\left[\left\{n\Sigma x^2-(\Sigma x)^2\right\}\left\{\dfrac{\Sigma y^2}{n}-\left(\dfrac{\Sigma y}{n}\right)^2\right\}\right]}}$$

Multiplying top and bottom by n we obtain

$$r = \frac{n\Sigma xy-\Sigma x\Sigma y}{\sqrt{[\{n\Sigma x^2-(\Sigma x)^2\}\{n\Sigma y^2-(\Sigma y)^2\}]}}$$

5.8.4 *The coefficient of determination* $= r^2$

The basic form of the coefficient is

$$\frac{\Sigma(\bar{y}-y')^2}{\Sigma(y-\bar{y})^2}$$

Taking the numerator first we get

$$\Sigma(\bar{y}-y')^2 = n\bar{y}^2+\Sigma y'^2-2\bar{y}\Sigma y'$$
$$= n\bar{y}^2+\Sigma(a+bx)^2-2\bar{y}\Sigma(a+bx)$$
$$= \frac{(\Sigma y)^2}{n}+na^2+b^2\Sigma x^2+2ab\Sigma x-2a\Sigma y-\frac{2b\Sigma x\Sigma y}{n}$$

Substituting $a = \dfrac{\Sigma y-b\Sigma x}{n}$ and multiplying by n we have

$$n\Sigma(\bar{y}-y')^2 = (\Sigma y)^2+(\Sigma y)^2+b^2(\Sigma x)^2-2b\Sigma x\Sigma y+nb^2\Sigma x^2+2b\Sigma x\Sigma y$$
$$-2b^2(\Sigma x)^2-2(\Sigma y)^2+2b\Sigma x\Sigma y-2b\Sigma x\Sigma y$$
$$= nb^2\Sigma x^2-b^2(\Sigma x)^2$$
$$= b^2[n\Sigma x^2-(\Sigma x)^2]$$

The denominator

$$\Sigma(y-\bar{y})^2 = \Sigma y^2 + \frac{(\Sigma y)^2}{n} - 2\frac{(\Sigma y)^2}{n}$$

Multiplying by n we obtain

$$n\Sigma(y-\bar{y})^2 = n\Sigma y^2 - (\Sigma y)^2$$

Thus

$$\frac{\Sigma(\bar{y}-y')^2}{\Sigma(y-\bar{y})^2} = \frac{b^2[n\Sigma x^2 - (\Sigma x)^2]}{n\Sigma y^2 - (\Sigma y)^2}$$

Letting $b = \dfrac{n\Sigma xy - \Sigma x\Sigma y}{n\Sigma x^2 - (\Sigma x)^2}$

gives

$$\frac{\left[\dfrac{n\Sigma xy - \Sigma x\Sigma y}{n\Sigma x^2 - (\Sigma x)^2}\right]^2 [n\Sigma x^2 - (\Sigma x)^2]}{n\Sigma y^2 - (\Sigma y)^2}$$

Cancelling and rearrangement yields

$$\frac{[n\Sigma xy - \Sigma x\Sigma y]^2}{[n\Sigma x^2 - (\Sigma x)^2][n\Sigma y^2 - (\Sigma y)^2]} = r^2$$

5.8.5 *To prove* $\Sigma(y'-\bar{y})^2 = a\Sigma x + b\Sigma xy - \dfrac{(\Sigma y)^2}{n}$

Expanding the left-hand side of $\Sigma(y'-\bar{y})^2 = a\Sigma x + b\Sigma xy - \dfrac{(\Sigma y)^2}{n}$ we get:

$$\Sigma(y'-\bar{y})^2 = \Sigma y'^2 + n\frac{(\Sigma y)^2}{n^2} - 2\Sigma y'\frac{\Sigma y}{n}$$

Now $\quad \Sigma y'^2 = \Sigma(a+bx)^2$

$$= \Sigma(a^2 + 2abx + b^2x^2)$$
$$= na^2 + 2ab\Sigma x + b^2\Sigma x^2$$
$$= a(na + b\Sigma x) + b(a\Sigma x + b\Sigma x^2)$$

But from the normal equations:

$$\Sigma y = na + b\Sigma x$$

and $\quad \Sigma xy = a\Sigma x + b\Sigma x^2$

Therefore $\quad \Sigma y'^2 = a\Sigma y + b\Sigma xy$

Because $\quad \Sigma y' = \Sigma y$

201 Two-variable linear regression and correlation

$$\Sigma(y'-\bar{y})^2 = a\Sigma y + b\Sigma xy + n\frac{(\Sigma y)^2}{n^2} - 2\frac{(\Sigma y)^2}{n}$$

$$= a\Sigma y + b\Sigma xy + \frac{(\Sigma y)^2}{n} - 2\frac{(\Sigma y)^2}{n}$$

$$= a\Sigma y + b\Sigma xy - \frac{(\Sigma y)^2}{n}$$

5.9   Examples

5.9.1   *Regression*

(a) The following data relates to the number of cigarettes per adult smoked in 1930 and the standardized death rate per million per year in 1952, for sixteen countries.

Calculate the regression of death rate on consumption of cigarettes, critically interpreting your answer.

Country	Cigarettes smoked per adult per annum	Death rate
England and Wales	1,378	461
Finland	1,662	433
Austria	960	380
Netherlands	632	276
Belgium	1,066	254
Switzerland	706	236
New Zealand	478	216
U.S.A.	1,296	202
Denmark	465	179
Australia	504	177
Canada	760	176
France	585	140
Italy	455	110
Sweden	388	89
Norway	359	77
Japan	723	40

(*British Medical Journal*, February 1959)

(b) A firm employing workers on piece-work rates of pay for packing boxes of biscuits, finds the following situation from an O. and M. study:

Number of boxes packed in ten minutes	Number of hours' work already completed in day
4·5	0·5
6·3	1·25
7·2	8·6
6·5	10·3
12·3	4·2
10·2	6·6
11·6	3·8
14·5	5·4
12·8	6·4
5·6	0·75

Find the linear regression of boxes packed on number of completed hours of work. Suggest why this may not be the most appropriate form of the relationship in this case (use a diagram to demonstrate this point).

(c) The following data show the average weekly household income and expenditure on durable household goods in the eleven standard regions of Great Britain (1961–3):

Region	Average weekly household income £	Average weekly expenditure on durable household goods £
North	18·55	1·15
East and West Ridings	20·00	1·10
North Midlands	20·10	1·15
East	20·40	1·20
London and south-east	22·75	1·40
South	22·55	1·00
South-west	19·95	1·05
Wales	19·05	0·95
Midlands	23·55	1·35
North-west	20·35	1·15
Scotland	19·50	1·05

(*Regional Statistics* 1965)

What would you expect the expenditure on durable household goods to be in a region having an average household income in 1965 of £25?

(d) A market research organization carries out a survey to investigate the reading habits of married women. The following results were obtained:

Number of books, periodicals etc. borrowed from library or purchased per month	Years married				
	0–3	3–9	9–15	15–21	21–33
20–15	12	—	—	3	12
14–12	7	—	1	8	7
11–9	2	—	5	10	3
8–6	1	6	6	2	—
5–3	—	5	1	—	—
2–0	—	3	—	—	—

Using appropriate regression techniques, how would you interpret the above data?

(e) In trying to evaluate the effectiveness of its advertising, a firm compiled the following information:

Year	Advertising expenditure (£000's)	Revenue (£00,000's)
1958	6	2·5
1959	12	3·8
1960	15	3·9
1961	15	4·2
1962	23	5·0
1963	24	4·8
1964	38	6·2
1965	42	6·0
1966	47	7·9

What is the regression of revenue on advertising expenditure? Would it be reasonable to estimate revenue in 1970 if advertising expenditure will be £65,000? If not, why not?

5.9.2   Correlation

(a) The following figures are index numbers of average weekly earnings and prices in the United Kingdom:

Average 1955 = 100

	1955	1956	1957	1958	1959	1960	1961	1962	1963	1964
Earnings	100	108	113	117	122	130	138	143	149	162
Prices	100	105	109	112	113	114	118	123	125	129

(Ministry of Labour)

Calculate the product-moment coefficient of correlation between prices and earnings, and explain the meaning of your answer.

(b) The marks of 195 children aged 15 in mathematics and Latin were as follows:

Maths	Latin 0–9	10–19	20–29	30–39	40–49	50–59	60–69	70–79	80–89	90–100
100–90	—	—	—	—	—	—	—	—	—	3
90–80	1	—	—	—	—	—	—	12	10	7
80–70	—	—	—	—	—	2	4	9	1	—
70–60	—	—	—	—	4	9	10	5	4	—
60–50	—	1	2	8	5	8	1	2	—	—
50–40	1	—	5	10	14	—	3	—	—	—
40–30	3	4	6	1	3	2	—	—	—	—
30–20	5	8	2	3	—	—	—	—	—	—
20–10	7	9	—	—	—	—	—	—	—	—
10–0	—	—	—	—	—	—	1	—	—	—

From the calculation of the coefficient of correlation, what might be inferred?

(c) The following table relates to the percentage voting in local electio
and the size of the county borough. Obtain a measure of the correlati
between them and comment.

| Number of electors | Percentage voting | | | | | |
	Up to 45	45– 50	50– 55	55– 60	60 and upwards	To
Up to 60,000	2	5	8	7	5	27
60–80,000	4	4	3	1	3	15
80–100,000	0	3	2	2	3	10
100–200,000	4	5	2	3	2	16
200–300,000	2	1	1	1	0	5
	12	18	16	14	13	73

(*Registrar-General's Annual Review*)

(d) The following shows the percentage of 17–19-year-olds in nir
counties in receipt of state scholarships and local education authorit
awards tenable at university:

	State scholarship (per cent)	Local education authority awards (per cent)
Durham	0·9	6·8
Staffordshire	1·2	5·6
Nottinghamshire	1·3	5·3
Herefordshire	1·3	5·2
Cumberland	1·8	6·6
Cornwall	1·9	4·7
Cheshire	3·1	12·1
Hertfordshire	3·4	8·7
Surrey	3·6	9·7

Calculate the correlation coefficient and carefully comment on an
interpret your results.

(e)

Industrial group	Gross income as percentage of gross assets	Dividend and interest payments as percentage of gross assets
Chemicals	14·0	4·6
Metal manufacture	9·7	3·7
Engineering	12·6	4·4
Electrical engineering	13·5	4·1
Shipbuilding	6·0	2·4
Vehicles	14·7	4·1
Textiles	12·3	4·6
Retailing	19·3	7·9
Clothing and footwear	14·2	5·6

What is the coefficient of correlation between the two variables above? Explain the meaning of the result in the context of the situation described.

Chapter 6
The analysis of time series

6.1 The nature of time series analysis

6.1.1 The techniques which have been developed up to this point provide us with a means of analysing a set of data which is assumed to have been collected at a moment in time. Let us turn now to a full discussion of the methods available for sifting the various factors which contribute to the changes over time in such variable series as imports and exports, sales, output, unemployment, and prices. If we can isolate and evaluate the importance of the main components which determine the value of, say, sales for a particular month, then we can project the series into the future to obtain a forecast.

It should be stressed from the beginning, however, that the mathematical and statistical determination and projection of the components of a time series does not necessarily produce accurate forecasts. The whole of the analysis which will be developed in this chapter rests upon the assumption of stable political, economic and social conditions during the historical period covered by the data being studied.

This does not imply a stagnant society: evolutionary changes in society can and will be taken into account. Abrupt and major changes, on the other hand, will produce marked jumps or declines in series which cannot be explained within the previous framework. The impact of the Second World War on the economy of countries like the United Kingdom or the United States was obviously considerable. Besides economic and technological implications, the war also produced a revolution in attitudes, opinions and motivations which was reflected socially and politically through changing government policies at home and abroad. Predictions and forecasts made in 1938 for the war years or post-war years must necessarily have been wrong because the analyst would have been working on the basis of evolutionary change and could not have been expected to foresee the exact time, nature or effect of the revolutionary changes which were about to occur. Nevertheless, once the war had started he might have used a knowledge of the effect of previous conflagrations to modify his predictions. Alternatively, the development of the new long-run pattern might be awaited so that the effect of the catastrophic disturbance is eliminated and a fresh evolutionary situation established.

It seems that although mathematical projections alone are largely valueless, they do provide a basis on which to build. Experience, judgement and intuition should be used to modify the anomalies which impersonal techniques produce. Only by combining an analysis of past events with the perspective of human judgement in evaluating sudden change will acceptable forecasts be made.

6.1.2   The reader's own experience should point to the major components which may be present in a time series and which must be isolated. The sales of motor-cars in the United Kingdom since the middle 1940s provides a good illustration. From an awareness of the increasing traffic congestion on the roads it is apparent that sales must have been increasing steadily during the past twenty years. As personal incomes have risen and technological innovation has produced a relative reduction in the price of cars, more and more people have been able to buy and own their own vehicles. The trend of car sales has been upwards.

Every series will possess a *secular trend* which results from the long-run influence of socio-politico-economic factors. The trend may be upwards (growth) or downwards (decline) but its major characteristic is that it continues in one direction or in a regular pattern for long periods.

Although the overall direction of car sales has been upwards, the reader will recognize that there have been regular periodic movements in the actual sales of cars associated with the time of the year. There tends to be a regular upsurge of sales in the spring when motorists see the prospect of week-end trips to the countryside or coast (or alternatively anticipate increases in taxation in the April national budget). Similarly, sales drop prior to the release of a new model and in the last few months of the year. The seasonal nature of the car trade is by no means unique. Employment and output in the construction and building industry depend upon the weather and thus the time of the year. The sales of greetings cards or fireworks are largely associated with public holidays and festivals and so will be subject to large seasonal variations. The list could be extended indefinitely. What does emerge is that the *seasonal component* of a time series is important and where appropriate must be given specific consideration.

Car sales have been subject to a second type of periodic movement which tends to be more irregular and longer in duration than the seasonal movements. These fluctuations have been associated with the health and the ill-health of the economy as a whole. Britain has been subject to a number of crises in the post-war decades, usually associated with her balance of payments disequilibrium. At these times, demand has been restricted by increased direct and indirect taxation and by curtailments on credit facilities. The ensuing reduction in general economic activity has frequently been most marked in the consumer durables industries, such as motor manufacturing. Sales have first dropped and then gradually picked up again until the next period of excessive demand, when further restrictions have had to be imposed.

These oscillations, although minor in nature compared with the business cycles of the nineteenth and early twentieth centuries, exhibit the same characteristics and features of their nineteenth-century counterparts. Although this is not the place to comment on the theories which have been put forward to explain the causes of cyclical fluctuations in market economies, we shall be interested to know whether there is a noticeable *cyclical component* in a series and whether at any particular time the series is on the down-turn or is rising from the trough of a recession. The main difficulty which will be encountered in practice is that only long-period sets of data will show cyclical fluctuations of any appreciable magnitude, and there are relatively few series for which lengthy and accurate records have been kept. In some of our examples, we shall therefore omit the cyclical component. In addition, we shall find that the assumptions made about the length of the cycles will themselves affect the pattern of cycles which emerges from the analysis.

Finally, returning to the car sales illustration, we shall find that we have not completely explained a particular level of sales in one month or quarter even after taking account of the trend, the seasonal component and the cyclical component. This *residual component* cannot be explained by any of the three concepts mentioned earlier. It is the result of purely random and irregular once-and-for-all events which are completely unpredictable. For instance, early spring weather may be particularly good, resulting in above normal car sales in one year, or general pessimism about the continuation of overtime and/or full employment may cause purchases to be deferred with the result that sales in the autumn of another year are abnormally low.

6.2 The decomposition of time series

6.2.1 If the secular trend is denoted by T, the seasonal component by S, the cyclical component by C and the residual component by R, we may establish two models which indicate the relationships between these in explaining the actual data, A. The additive model, which is the simpler to use arithmetically, assumes that the actual data is the sum of the four separate effects, i.e.

$$A = T+S+C+R$$

The multiplicative model, on the other hand, suggests that the actual data are the product of the effects of the trend and the seasonal, cyclical and residual components, i.e.

$$A = T \times S \times C \times R$$

6.2.2 The additive model assumes that the effect of the seasons, the cycles and the residual component are equal in absolute terms throughout the period being studied. While this is probably an adequate assumption when short periods are involved, or where the rate of growth or decline in the trend is small, there must be doubts when one considers a long time-span and a marked growth rate. In these circumstances the amplitude of the cycles is more likely to be proportionate to the trend, and the amplitude of the seasonal component proportionate to the stage in the life of the cycle. Put another way, we may expect that the oscillations in the data from peak to peak in the cycles will increase in absolute terms as the trend increases, and that the magnitude of the seasonal fluctuations will be greater near the peak of a cycle than in the trough. It is for this reason that the multiplicative model has found more favour over recent years.

6.2.3 Either of these models may be used to effect the decomposition of the time series. The first step will usually be to estimate the trend (see 6.3) and then to eliminate the trend for each time period (day, week, month or quarter) from the actual data by subtraction or division, giving a detrended series which expresses the effect of the seasons, the cycles and the residual component.

Detrended series

Additive model

$$A-T = S+C+R$$

Multiplicative model

$$\frac{A}{T} = S \times C \times R$$

The detrended series is then averaged day by day, month by month, or quarter by quarter to produce an estimate of the seasonal component,

S (assuming that S is predominant and that $C+R = 0$ and $C \times R = 1$); a deseasonalized or seasonally adjusted series can now be obtained.

Seasonally adjusted series

Additive model

$$A - S = T + C + R$$

Multiplicative model

$$\frac{A}{S} = T \times C \times R$$

The elimination of the seasonal component can by itself be most valuable to the economist or businessman, who wishes to know whether a change from month to month or quarter to quarter reflects an increase or decrease in, for instance, export or sales performance. Looking at an unadjusted series he will find difficulty in deciding, for the figures will be disguised by the seasonal component.

The next step is to obtain a detrended, deseasonalized series by either subtracting S or T from one of the expressions above (additive model), or dividing one of them by S or T (multiplicative model).

Detrended, deseasonalized series

Additive model

$$A - T - S = C + R$$

Multiplicative model

$$\frac{A}{T \times S} = C \times R$$

Finally, C can be found by smoothing the joint C and R component and is, as before, eliminated.

Residual component

Additive model

$$A - T - S - C = R$$

Multiplicative model

$$\frac{A}{T \times S \times C} = R$$

6.2.4   Although the general method of decomposition has included the four possible components which make up a time series, we should again emphasize that it is not an invariable rule for all four to be present. If annual data are confronted, there can be no seasonal component. Similarly, if short periods of time are involved, the cyclical component can be ignored. In both cases one of the steps in the decomposition of the time series outlined above may be omitted.

6.3   The trend

6.3.1   There are a number of methods, each valuable in its place, by which the secular trend of a series may be estimated. Which one will be employed in practice depends upon the nature of the data and the purpose of the analysis. Briefly, the methods are as follows:

(i) Freehand curve fitting
(ii) The semi-average method
(iii) Simple moving averages
(iv) Exponentially weighted moving averages
(v) Mathematical trend curves

In this chapter we will cover the first three of these, together with the linear trend, which is the simplest system from category (v). The remaining methods are dealt with in *Statistics for the Social Scientist: 2 Applied Statistics.*

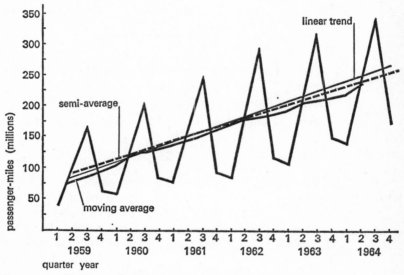

Fig. 64   Passenger-miles flown: actual data and trend estimates.

6.3.2   As in regression analysis a freehand curve may be superimposed over the histogram to obtain an approximate estimate of the trend. However, the same limitation of subjectivity which occurred in our earlier attempts to fit the line of best fit to the scatter diagram now arises: no two people using this method will estimate the same two sets of trend values, and the technique should in consequence be employed only sparingly and in those cases where the graph of the original data already conforms to some well-defined pattern. If under these circumstances the pattern happens to be linear, we may utilize the second of our methods, the semi-average trend, which enables a much better estimate of the trend to be made. To illustrate this, let us use the data shown below and the graph drawn from it (fig. 64).

Passenger-miles flown by domestic services of United Kingdom airlines (millions of passenger-miles)

Year	Quarter First (Jan.–Mar.)	Second (Apr.–June)	Third (July–Sept.)	Fourth (Oct.–Dec.)
1959	43·2	96·4	161·2	63·4
1960	58·0	128·4	207·5	83·6
1961	77·4	166·3	251·9	104·4
1962	98·9	191·0	287·4	123·2
1963	113·4	228·8	316·2	155·7
1964	151·5	254·8	357·1	180·3

(*Monthly Digest of Statistics*)

This series is now split chronologically into halves, and the mean of the quarterly figures in each half calculated. In this case the mean number of passenger-miles flown in the twelve quarters from January 1959 to December 1961 is $\dfrac{1441·7}{12} = 120·14$, while from January 1962 to December 1964 the mean is $\dfrac{2458·3}{12} = 204·86$. These mean values are plotted at the centre of the time span to which they relate and the points joined up by a straight line. The first value of 120·14 therefore goes between the second and third quarters of 1960 and the second, 204·86, between the second and third quarters of 1963. The trend is shown graphically in fig. 64 and the trend values, interpolated from the graph, are given below:

Passenger-miles flown by domestic services of United Kingdom airlines (millions of passenger miles)

Semi-average trend

Year	Quarter First (Jan.–Mar.)	Second (Apr.–June)	Third (July–Sept.)	Fourth (Oct.–Dec.)
1959	81·31	88·37	95·43	102·49
1960	109·55	116·61	123·67	130·73
1961	137·79	144·85	151·91	158·97
1962	166·03	173·09	180·15	187·21
1963	194·27	201·33	208·39	215·45
1964	222·35	229·57	236·63	243·69

6.3.3 *Simple moving averages*

6.3.3.1 A moving average estimate of the trend can be produced in the following manner. The first n observations are summed and divided by n, enabling the trend value for the middle period of the n observations to be estimated by this mean. The first observation in the series is then dropped and the $n+1$ observation included: the new average is the trend value for the next time period. Put algebraically the trend value for period t using a five-period moving average is

$$\text{M.A.}_t = \tfrac{1}{5}(A_{t-2}+A_{t-1}+A_t+A_{t+1}+A_{t+2})$$

while for period $t+1$ it is

$$\text{M.A.}_{t+1} = \tfrac{1}{5}(A_{t-1}+A_t+A_{t+1}+A_{t+2}+A_{t+3})$$

A simple application of this moving average technique is shown below.

The annual data for the output of steel from the United Kingdom industry are obviously subject to cyclical fluctuations (see fig. 65) and the calculation and plotting of the five-year moving average does little to eradicate these cycles. If the periodicity of the cycles were constant, so that successive peaks and troughs were exactly the same number (m)

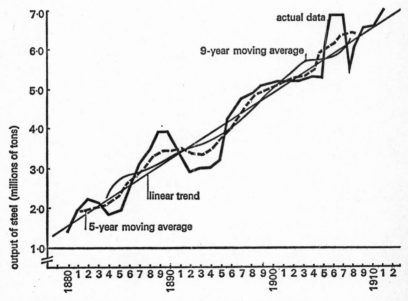

Fig. 65 United Kingdom steel output: actual data and trend estimates.

Output of steel ingots and castings in the United Kingdom (millions of tons)

Moving average trend

Year	Actual data	Sum of 5 years	5-year moving average	Sum of 9 years	9-year moving average
1880	1·3				
1881	1·8				
1882	2·1		1·80		
1883	2·0		1·92		
1884	1·8	9·0	2·02		2·16
1885	1·9	9·6	2·20		2·42
1886	2·3	10·1	2·46		2·62
1887	3·0	11·0	2·82		2·74
1888	3·3	12·3	3·16	19·5	2·84
1889	3·6	14·1	3·34	21·8	2·98
1890	3·6	15·8	3·32	23·6	3·11
1891	3·2	16·7	3·26	24·7	3·22
1892	2·9	16·6	3·16	25·6	3·34
1893	3·0	16·3	3·10	26·8	3·48
1894	3·1	15·8	3·28	28·0	3·59
1895	3·3	15·5	3·60	29·0	3·72
1896	4·1	16·4	3·92	30·1	3·91
1897	4·5	18·0	4·26	31·3	4·13
1898	4·6	19·6	4·58	32·3	4·34
1899	4·8	21·3	4·74	33·5	4·56
1900	4·9	22·9	4·82	35·2	4·74
1901	4·9	23·7	4·90	37·2	4·93
1902	4·9	24·1	4·94	39·1	5·16
1903	5·0	24·5	5·12	41·0	5·37
1904	5·0	24·7	5·44	42·7	5·42
1905	5·8	25·6	5·76	44·4	5·53
1906	6·5	27·2	5·82	46·4	5·70
1907	6·5	28·8	6·00	48·3	5·88
1908	5·3	29·1	6·12	48·8	6·08
1909	5·9	30·0	6·12	49·8	
1910	6·4	30·6	6·18	51·3	
1911	6·5	30·6		52·9	
1912	6·8	30·9		54·7	

of years apart then an m-point moving average would suffice to produce the trend estimate. As such a perfect situation will rarely be encountered in the real world, we have to compromise. In this example the peaks of the cycles appear to lie between seven and ten years apart. A nine-year moving average was therefore plotted, and although there are still slight oscillatory movements it is evident that this curve does largely reflect the long-term growth of the steel industry in the United Kingdom.

6.3.3.2   Certain aspects of the method of moving averages must now be considered. From a computational point of view, it should be noticed that in the table on page 216 the sums of five and nine years have been placed alongside the last of the years included. This avoids the danger of misreading which figures are to be covered, and also enables a simple but very useful calculation aid to be employed by the analyst who does not possess a calculating machine; this is shown schematically below in connexion with the calculation of the five-year moving average.

Year	Actual data	Sum of five years
1880	1·3	
1881	1·8	
1882	2·1	
1883	2·0	
1884	1·8	9·0
1885	1·9	9·0+(1·9 – 1·3) = 9·6
1886	2·3	9·6+(2·3 – 1·8) = 10·1

After the first five figures have been added, i.e. $1·3+1·8+2·1+2·0+1·8 = 9·0$, we prepare a slip of paper with arrows worked at six-year intervals. These arrows therefore show which figure is to be subtracted, and which is to be added, to produce the next 'sum of five years'. Now instead of adding $1·8+2·1+2·0+1·8+1·9 = 9·6$, we find the difference between the figure which is omitted and the new figure which is to be included, and add this difference to the previous 'sum of five years'. In the first case, the omitted figure is 1·3, the newly included figure is 1·9: the difference is therefore $+0·6$, so that the new entry in the second column is $9+0·6 = 9·6$. The slip of paper may then be moved down one line to indicate the next omission and inclusion. The method is very much faster than the repeated summing of five figures and the marked slip of paper ensures that the correct data are being utilized at each step.

Once the moving five-yearly totals have been calculated, division by five gives the moving average which is centred in the middle of the five years to which it relates. The moving average for the years 1880–4 is written against 1882 and so on down the table.

6.3.3.3   The two principal disadvantages of this method may already be apparent to the reader. The method prohibits the calculation of trend values for the first and last $\frac{(n-1)}{2}$ time periods. Thus for the five-year moving average we have no figures for the first two and the last two years, while for the nine-year moving average we lose a total of eight years, four at each end of the series. Similarly, the moving average curve has no mathematical regularity of form and can only be extended for the purpose of projection by a freehand technique. Notwithstanding these objections (and a third, which is soon to be encountered), the simple moving average method has found wide acceptance amongst those social scientists who are more interested in analysing the past than in predicting the future.

6.3.3.4   Our discussion has so far concentrated on odd number-of-point moving averages because of the ease with which we may centre the results. If, however, we are to deal with monthly or quarterly data showing a marked seasonal pattern we must keep units of one year intact, so that each trend estimate is an average of the twelve different months or four different quarters of the year. We will have to use twelve-monthly or four-quarterly moving averages, and resolve in some convenient way the difficulty of centering on a specific time period. There is also a second aspect to this requirement of using an averaging procedure which spans only one year. It has already been established that cycles within a time series may cover as many as twenty-five years. It is hardly likely, therefore, that twelve-monthly or four-quarterly smoothing will remove oscillations in the cyclical pattern. What will emerge from the use of the short-span moving average is a series which jointly estimates the trend and the cycles.

In these circumstances, the steps in the decomposition of the time series will have to be modified. After the calculation of the moving average a detrended-decycled series is determined.

Detrended-decycled series

Additive model

$$A - (T + C) = S + R$$

Multiplicative model

$$\frac{A}{T \times C} = S \times R$$

The seasonal component, S, is then isolated as before to obtain the seasonally adjusted series.

Seasonally adjusted series

Additive model

Multiplicative model

$$A-S = T+C+R$$

$$\frac{A}{S} = T \times C \times R$$

At this stage the trend is estimated, using either a more appropriate and longer-run moving average or a mathematical trend curve and then eliminated to give the cyclical-residual movements.

Detrended-deseasonalized series

Additive model

Multiplicative model

$$A-S-T = C+R$$

$$\frac{A}{S \times T} = C \times R$$

The smoothing to obtain the cyclical component may then be carried out, and the residual component estimated. The main point to remember is that the answers obtained from this alternative procedure should be the same as for the one discussed earlier.

6.3.3.5   To illustrate how the difficulty of centering is overcome, the data given in 6.3.2 will be used. The sums of four quarters are found, and these are then summed in moving pairs to give the sums of eight quarters (see table on page 220).

Admittedly the entries in the column headed 'Sum of eight quarters' span five quarters chronologically, but each is made up of two first quarters, two second quarters, two third quarters and two fourth quarters. For instance, the first number in this column is 743·2, which is the sum of:

First quarter 1959 =   43·2
2 × Second quarter 1959 = 2 × 96·4
2 ×   Third quarter 1959 = 2 × 161·2
2 × Fourth quarter 1959 = 2 × 63·4
First quarter 1960 =   58·0
Eight quarters in all = $\overline{743·2}$

The integrity of the quarterly structure of the year has been maintained, yet because chronologically we span five quarters, the average can be placed against the middle quarter (in the first case the third quarter of 1959) instead of being located between two quarters as would happen if a simple four-point average were used. Not only is this a convenient method of centering the moving average values: it also provides a more

Passenger miles flown by domestic services of United Kingdom airlines (millions of passenger-miles)

Moving average trend (or trend + cycles)

Year	Quarter	Actual data	Sum of four quarters	Sum of eight quarters	Eight-point moving average
1959	1	43·2			
	2	96·4			
	3	161·2			92·90
	4	63·4	364·2		98·75
1960	1	58·0	379·0	743·2	108·54
	2	128·4	411·0	790·0	116·85
	3	207·5	457·3	868·3	121·80
	4	83·6	477·5	934·8	128·96
1961	1	77·4	496·9	974·4	139·25
	2	166·3	534·8	1031·7	147·40
	3	251·9	579·2	1114·0	152·69
	4	104·4	600·0	1179·2	158·46
1962	1	98·9	621·5	1221·5	165·99
	2	191·0	646·2	1267·7	172·78
	3	287·4	681·7	1327·9	176·94
	4	123·2	700·5	1382·2	183·48
1963	1	113·4	715·0	1415·5	191·80
	2	228·8	752·8	1467·8	199·46
	3	316·2	781·6	1534·4	208·29
	4	155·7	814·1	1595·7	216·30
1964	1	151·5	852·2	1666·3	224·66
	2	254·8	878·2	1730·4	232·85
	3	357·1	919·1	1797·3	
	4	180·3	943·7	1862·8	

efficient smoothing, because double the weight is given to the three middle quarterly figures utilized, i.e.

$$\text{M.A.}_t = \tfrac{1}{8}(A_{t-2}+2A_{t-1}+2A_t+2A_{t+1}+A_{t+2})$$

The moving average trend has been plotted in fig. 64 to give a comparison with the other methods.

6.3.4 Linear trends

6.3.4.1 The final method of trend estimation which will be discussed in this

Output of steel ingots and castings in the United Kingdom (millions of tons)

Linear trend

Year	Output y	Time units One year round 1880 x	round 1896	x^2	xy	Trend $y_t = 4.079 + 0.163\,x_t$
1880	1·3	0	−16	256	−20·8	1·471
1881	1·8	1	−15	225	−27·0	1·634
1882	2·1	2	−14	196	−29·4	1·797
1883	2·0	3	−13	169	−26·0	1·960
1884	1·8	4	−12	144	−21·6	2·123
1885	1·9	5	−11	121	−20·9	2·286
1886	2·3	6	−10	100	−23·0	2·449
1887	3·0	7	−9	81	−27·0	2·612
1888	3·3	8	−8	64	−26·4	2·775
1889	3·6	9	−7	49	−25·2	2·938
1890	3·6	10	−6	36	−21·6	3·101
1891	3·2	11	−5	25	−16·0	3·264
1892	2·9	12	−4	16	−11·6	3·427
1893	3·0	13	−3	9	−9·0	3·590
1894	3·1	14	−2	4	−6·2	3·753
1895	3·3	15	−1	1	−3·3	3·916
1896	4·1	16	0	0	0	4·079
1897	4·5	17	1	1	4·5	4·242
1898	4·6	18	2	4	9·2	4·405
1899	4·8	19	3	9	14·4	4·568
1900	4·9	20	4	16	19·6	4·731
1901	4·9	21	5	25	24·5	4·894
1902	4·9	22	6	36	29·4	5·057
1903	5·0	23	7	49	35·0	5·220
1904	5·0	24	8	64	40·0	5·383
1905	5·8	25	9	81	52·2	5·546
1906	6·5	26	10	100	65·0	5·709
1907	6·5	27	11	121	71·5	5·872
1908	5·3	28	12	144	63·6	6·035
1909	5·9	29	13	169	76·7	6·198
1910	6·4	30	14	196	89·6	6·361
1911	6·5	31	15	225	97·5	6·524
1912	6·8	32	16	256	108·8	6·687
	134·6		0	2,992	486·5	

chapter follows logically from chapter 5, where we derived expressions for fitting a straight line to the relationship between two variables. The least-squares method may now be employed, with a little modification and some simplification, to fit a linear trend to a time series. Time now takes the place of the x variable, while the single variable magnitude in the series is denoted by y. Obviously we cannot proceed with time periods such as '1880' or '1881'. We must convert years, quarters and months into time units round some arbitrary origin.

This has been done in the table above for the data on the output of steel between 1880 and 1912. First we have shown the years deviating round 1880; we could employ these time units in the calculations and find b and a using either the simultaneous equations method (see 5.3.4) or the formula method (see 5.3.6). It is better, however, to centre the origin on 1896 so that the sum of the time units equals zero (i.e. $\Sigma x = 0$). The normal equations then reduce to

$$\Sigma y = na \qquad \text{(i)}$$
$$\Sigma xy = b\Sigma x^2 \qquad \text{(ii)}$$

so that

$$a = \frac{\Sigma y}{n} \text{ and } b = \frac{\Sigma xy}{\Sigma x^2}$$

In this example

$$a = \frac{134 \cdot 6}{33} = 4 \cdot 079$$

$$b = \frac{486 \cdot 5}{2992} = 0 \cdot 163$$

so that the trend of steel output may be stated as

$$y_t = 4 \cdot 079 + 0 \cdot 163 x_t$$

(Origin 1896; unit one year.)

By substituting $x_t = -16, -15 \ldots 15, 16$, we may calculate the individual trend values as shown. Inspection of fig. 65 indicates that they are not markedly different from those obtained from the nine-year moving average.

6.3.4.2   Fitting a linear trend to the passenger-miles data differs from the above in only one detail. Whereas we were confronted with an odd number of years in the steel output series, we now face an even number of quarters. This means that we are unable to centre the time periods on one particular quarter. If, however, we express the time units in half quarters ($1\frac{1}{2}$ months) and centre the series half-way between the middle two quarters (i.e. the fourth quarter of 1961 and the first quarter of 1962) we overcome this difficulty.

Passenger-miles flown by domestic services of United Kingdom airlines (millions of passenger-miles)

Linear trend

Year	Quarter	Actual data y	Time units round Q.1 of 1959	round Q.4 of 1961 and Q.1 of 1962 x	x^2	xy	Trend $y_t = 162.5 + 3.707x$
1959	1	43·2	0	−23	529	−993·6	77·239
	2	96·4	2	−21	441	−2024·4	84·653
	3	161·2	4	−19	361	−3062·8	92·067
	4	63·4	6	−17	289	−1077·8	99·481
1960	1	58·0	8	−15	225	−870·0	106·895
	2	128·4	10	−13	169	−1669·2	114·309
	3	207·5	12	−11	121	−2282·5	121·723
	4	83·6	14	−9	81	−752·4	129·137
1961	1	77·4	16	−7	49	−541·8	136·551
	2	166·3	18	−5	25	−831·5	143·965
	3	251·9	20	−3	9	−755·7	151·379
	4	104·4	22	−1	1	−104·4	158·793
1962	1	98·9	24	1	1	−98·9	166·207
	2	191·0	26	3	9	573·0	173·621
	3	287·4	28	5	25	1437·0	181·035
	4	123·2	30	7	49	862·4	188·449
1963	1	113·4	32	9	81	1020·6	195·863
	2	228·8	34	11	121	2516·8	203·277
	3	316·2	36	13	169	4110·6	210·691
	4	155·7	38	15	225	2335·5	218·105
1964	1	151·5	40	17	289	2575·5	225·519
	2	254·8	42	19	361	4841·2	232·933
	3	357·1	44	21	441	7499·1	240·347
	4	180·3	46	23	529	4146·9	247·761
		3900·0		0	4600	17051·4	

From the tabulated calculations

$$a = \frac{3900}{24} = 162.5$$

$$b = \frac{17051.4}{4600} = 3.707$$

and the trend is

$$y_t = 162.5 + 3.707x_t$$

(Origin midway between fourth quarter 1961 and first quarter 1962; unit half quarters or $1\frac{1}{2}$ months.)

This is the third estimate of the trend shown in fig. 64.

6.3.4.3   Fitting a mathematical trend curve to time series data has obvious applications in forecasting, because the curve may easily be projected forwards as the basis of a forecast. The other commendation of this technique is that trend values for the whole series are made available, thus permitting a complete analysis of historical data to be undertaken. It seems that so long as there are justifications for fitting some form of mathematical curve to the data under review this method provides a valuable contribution to our analysis. Other forms of mathematical trend (non-linear) are discussed in *Statistics for the Social Scientist: 2 Applied Statistics.*

6.4   The additive model: seasonal analysis

6.4.1   The remainder of the time series analysis may now be undertaken. To demonstrate this we shall use the moving average trend as the first step in the estimation of the seasonal component for the additive model; in the case of the multiplicative model, the linear trend will be employed. In the former case we have in reality estimated T and C, but as the series covers only six years and because there is no noticeable cyclical pattern present, the simplified model

$$A = T + S + R$$

will be more relevant.

6.4.2   After tabulating the actual data and the trend values we subtract the latter from the former to obtain the detrended series, which reflects both the seasonal and the residual component. To eliminate the residual component we average all the detrended figures for the first quarter, for the second quarter, for the third quarter, and for the fourth quarter.

It will be noticed that the sum of the detrended values does not equal zero. If the trend were completely efficient, however, and if the omitted quarters had been included, the effect of the second and third quarters (with greatly increased holiday season air traffic) would be exactly offset by the poor figures in the first and fourth quarters. Because $(-331 \cdot 04 + (99 \cdot 96) + (471 \cdot 58) + (-255 \cdot 65) = -15 \cdot 15$ and not zero we should adjust each of these seasonal totals by subtracting $\dfrac{-15 \cdot 15}{4} = -3 \cdot 79$

This gives the corrected figures shown, which are then divided by 5 to produce the seasonal component for each quarter. We find that the effect of the first quarter of the year is to reduce the number of passenger miles flown by $65 \cdot 45$ million, the effect of the second quarter is to increase passenger-miles flown by $20 \cdot 75$ million, and the effect of the

Passenger-miles flown by domestic services of United Kingdom airlines
(millions of passenger-miles)

Seasonal analysis using the additive model

Year	Quarter	A	Moving average trend T	Detrended series A−T = S+R	Seasonal component S	Seasonally adjusted series A−S = T+R	Residual component A−S−T = R
1959	1	43·2			−65·45	108·65	
	2	96·4			20·75	75·65	
	3	161·2	92·90	68·30	95·07	66·13	−26·77
	4	63·4	98·75	−35·35	−50·37	113·77	15·02
1960	1	58·0	108·54	−50·54	−65·45	123·45	14·91
	2	128·4	116·85	11·55	20·75	108·65	−9·20
	3	207·5	121·80	85·70	95·07	112·43	−9·37
	4	83·6	128·96	−45·36	−50·37	113·97	5·01
1961	1	77·4	139·25	−61·85	−65·45	142·85	3·60
	2	166·3	147·40	18·90	20·75	145·55	−1·85
	3	251·9	152·69	99·21	95·07	156·83	4·14
	4	104·4	158·46	−54·06	−50·37	154·77	−3·69
1962	1	98·9	165·99	−67·09	−65·45	164·35	−1·64
	2	191·0	172·78	18·22	20·75	170·25	−2·53
	3	287·4	176·94	110·46	95·07	192·33	15·39
	4	123·2	183·48	−60·28	−50·37	173·57	−9·91
1963	1	113·4	191·80	−78·40	−65·45	178·85	−12·95
	2	228·8	199·46	29·34	20·75	208·05	8·59
	3	316·2	208·29	107·91	95·07	221·13	12·84
	4	155·7	216·30	−60·60	−50·37	206·07	−10·23
1964	1	151·5	224·66	−73·16	−65·45	216·95	−7·71
	2	254·8	232·85	21·95	20·75	234·05	1·20
	3	357·1			95·07	262·03	
	4	180·3			−50·37	230·67	

Detrended series

Year	First quarter	Second quarter	Third quarter	Fourth quarter
1959			68·30	−35·35
1960	−50·54	11·55	85·70	−45·36
1961	−61·85	18·90	99·21	−54·06
1962	−67·09	18·22	110·46	−60·28
1963	−78·40	29·34	107·91	−60·60
1964	−73·16	21·95		
Total	−331·04	99·96	471·58	−255·65
Corrected −(−3·79)	−327·25	103·75	475·37	−251·86
Average (÷5)	−65·45	20·75	95·07	−50·37

225   The analysis of time series

third and fourth quarters to increase and decrease traffic by 95·07 and 50·37 millions of passenger-miles respectively. The tabulation and subtraction of the quarterly seasonal components from the actual data yield the seasonally adjusted series. Finally, the trend values are subtracted from the seasonally adjusted figures, or alternatively the seasonal component is subtracted from the detrended series to enable an estimate of the residual movements to be made. As a check that the calculations are correct any quarter can be selected to see whether $T+S+R = A$. For instance, in the third quarter of 1960, $T = 121·80$, $S = 95·07$, and $R = -9·37$. Summing gives 207·5, which is correct.

6.4.3   As a useful means of indicating the procedure which has been adopted in the analysis, graphs of the trend and seasonally adjusted series should be superimposed over the original historigram. This has been done in fig. 66.

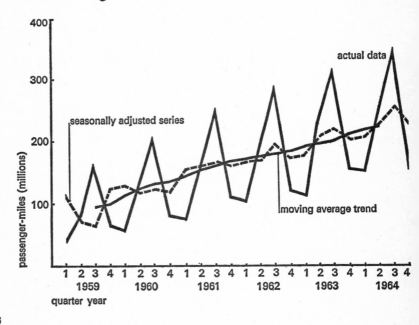

Fig. 66

It can be seen that the trend which removes the influence of the seasons and the residual component ($T = A-S-R$) is subject to the least fluctuation, as should be expected. The seasonally adjusted series ($A-S = T+R$) is merely the trend plus the residual component, so that it shows only minor oscillations round the trend. For any quarter the detrended figure can be obtained by interpolation from fig. 66 as the vertical distance between the actual and the trend points. Similarly the residual component is the vertical distance between the seasonally adjusted series and the trend.

6.4.4   As an additional aid to interpretation, the detrended series and the seasonal component may be plotted as in fig. 67.

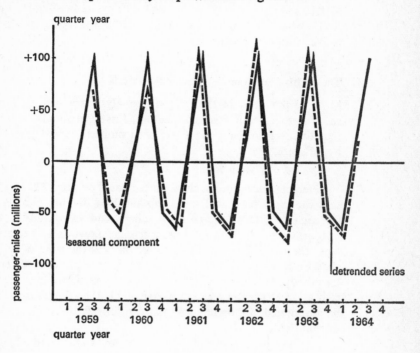

Fig. 67

Although the model has assumed that the effect of the seasons is constant over the whole period, it emerges that this is incorrect. The regularity in the detrended series suggests that this largely represents the seasonal component, but the detrended series is seen to be increasing in magnitude between 1959 and 1964, while the seasonal pattern is fixed in absolute terms. The residual component, which graphically is the vertical distance between the two curves in fig. 67, must therefore contain some of the seasonal component. This is also shown by the marked regularity of the residual component in the tabulation; over the first quarters the residual component is positive when the seasonal component is negative, and vice versa. This shows that the estimated magnitude of the seasonal fluctuations is too great for these quarters. In the last year of the series, on the other hand, the residual component has the same sign as the seasonal component which suggests that the averaging procedure has produced an understatement of the seasonal fluctuations at this stage. It seems that the additive model using a moving average trend suffers from certain marked deficiencies when applied to this set of data and the multiplicative model, with its stress on the proportionate relationship between the components within the series, should therefore prove more appropriate.

6.5 The multiplicative model: seasonal analysis

6.5.1 Most of the steps in the following analysis are straightforward, but certain aspects will need clarification. Firstly, it should be appreciated that the cyclical component has been omitted once again, so that we hypothesize

$$A = T \times S \times R.$$

Referring once more to the data tabulated on page 229, it will be seen that the ratio detrended series, obtained by dividing the actual figure for each quarter by the corresponding trend value, indicates by what proportion the number of passenger-miles flown in any quarter differs from the trend. For the first quarter of 1959 the ratio of $A/T = 0.559$ shows that the observed value of 43·2 is 44·1 per cent lower than the trend value of 77·239. Similarly $A/T = 1.139$ shows that A is 13·9 per cent above T for the second quarter of 1959. Evidently these ratios can be interpreted in the same way as the price relatives and index numbers in chapter 4.

The averaging procedure which removes the residual component follows the same pattern as before except that the adjustment to the totals which corrects for rounding errors is achieved by dividing 24 by the sum of the detrended ratios, i.e. 23·984, and multiplying the resulting quotient by the unadjusted totals. In this case the answer is $\frac{24}{23 \cdot 984} = 1.0006671$. The detrended ratios may be expected to equal 24 because if there were no seasonal component and no residual variation the original data would be the same as the trend, i.e. $A/T = 1.00$ for each quarter; thus A/T for quarterly data over six years would equal 24. But the seasonal component for the good quarters should be exactly offset by that for the bad quarters, so A/T should still equal 24.

After the seasonally adjusted series has been calculated, the residual

Passenger-miles flown by domestic services of United Kingdom airlines (millions of passenger-miles)

Seasonal analysis using multiplicative model

Year	Quarter	Actual data A	Linear trend T	Ratio detrended series A/T = S × R	Seasonal index S	Seasonally adjusted series A/S = T × R	Residual index A/(T × S) = R
1959	1	43·2	77·239	0·559	0·586	73·720	0·954
	2	96·4	84·653	1·139	1·124	85·765	1·013
	3	161·2	92·067	1·751	1·617	99·691	1·083
	4	63·4	99·481	0·637	0·673	94·205	0·947
1960	1	58·0	106·895	0·543	0·586	98·976	0·927
	2	128·4	114·309	1·123	1·124	114·235	0·999
	3	207·5	121·723	1·705	1·617	128·324	1·054
	4	83·6	129·137	0·647	0·673	124·220	0·961
1961	1	77·4	136·551	0·567	0·586	132·082	0·968
	2	166·3	143·965	1·155	1·124	147·954	1·028
	3	251·9	151·379	1·664	1·617	155·782	1·029
	4	104·4	158·793	0·657	0·673	155·126	0·976
1962	1	98·9	166·207	0·595	0·586	168·771	1·015
	2	191·0	173·621	1·100	1·124	169·929	0·979
	3	287·4	181·035	1·588	1·617	177·737	0·981
	4	123·2	188·449	0·654	0·673	183·061	0·972
1963	1	113·4	195·863	0·579	0·586	193·515	0·988
	2	228·8	203·277	1·126	1·124	203·559	1·002
	3	316·2	210·691	1·501	1·617	195·547	0·928
	4	155·7	218·105	0·714	0·673	231·352	1·061
1964	1	151·5	225·519	0·672	0·586	258·532	1·147
	2	254·8	232·933	1·094	1·124	226·690	0·973
	3	357·1	240·347	1·486	1·617	220·841	0·919
	4	180·3	247·761	0·728	0·673	267·905	1·082

Ratio detrended series

Year	First quarter	Second quarter	Third quarter	Fourth quarter
1959	0·559	1·139	1·751	0·637
1960	0·543	1·123	1·705	0·647
1961	0·567	1·155	1·664	0·657
1962	0·595	1·100	1·588	0·654
1963	0·579	1·126	1·501	0·714
1964	0·672	1·094	1·486	0·728
Total	3·515	6·737	9·695	4·037
Corrected (×24/23·984)	3·517	6·742	9·701	4·040
Average (÷ 6)	0·586	1·124	1·617	0·673

ratio is obtained either by dividing these figures by the trend values or by dividing the ratio detrended series by the respective seasonal indices. Having completed the analysis a check may again be applied by substituting for any quarter the values of T, S and R. For the third quarter of

1960, $T = 121 \cdot 723$, $S = 1 \cdot 617$ and $R = 1 \cdot 054$, so that $T \times S \times R$ = $121 \cdot 723 \times 1 \cdot 617 \times 1 \cdot 054 = 207 \cdot 455$ which is correct (allowing for rounding errors).

6.5.2   Graphs of the various series might be produced as before, and the reader is recommended to take this step. He will find on this occasion that many of the deficiencies found with the use of the additive model have disappeared. The increasing absolute magnitude of the seasonal fluctuations has been taken into account and the residual index seems to be completely random and unassociated with the seasonal index.

6.6   The multiplicative model: cyclical analysis

6.6.1   The steel output data (page 216) possessed a noticeable cyclical pattern, although being an annual record it lacked any seasonal component. The cyclical component will now be analysed using the model

$$A = T \times C \times R$$

and employing the linear trend calculated in 6.3.8. The preliminary calculations to obtain a detrended series do not differ from those outlined

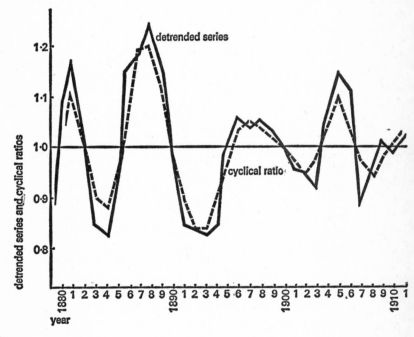

Fig. 68

Output of steel ingots and castings in the United Kingdom (millions of tons)

Year	Actual data A	Linear trend T	Ratio detrended series $A/T = C \times R$	Sum of $3(C \times R)$	Cyclical ratio C	Residual ratio $A/(T \times C) = R$
1880	1·3	1·471	0·884			
1881	1·8	1·634	1·102		1·052	1·048
1882	2·1	1·797	1·169	3·155	1·097	1·066
1883	2·0	1·960	1·020	3·291	1·012	1·008
1884	1·8	2·123	0·848	3·037	0·900	0·942
1885	1·9	2·286	0·831	2·699	0·873	0·952
1886	2·3	2·449	0·939	2·618	0·973	0·965
1887	3·0	2·612	1·149	2·919	1·092	1·052
1888	3·3	2·775	1·189	3·277	1·188	1·001
1889	3·6	2·938	1·225	3·563	1·192	1·028
1890	3·6	3·101	1·161	3·575	1·122	1·035
1891	3·2	3·264	0·980	3·366	0·996	0·984
1892	2·9	3·427	0·846	2·987	0·887	0·954
1893	3·0	3·590	0·836	2·662	0·836	1·000
1894	3·1	3·753	0·826	2·508	0·835	0·989
1895	3·3	3·916	0·843	2·505	0·891	0·946
1896	4·1	4·079	1·005	2·674	0·970	1·036
1897	4·5	4·242	1·061	2·909	1·037	1·023
1898	4·6	4·405	1·044	3·110	1·052	0·992
1899	4·8	4·568	1·051	3·156	1·044	1·007
1900	4·9	4·731	1·036	3·131	1·029	1·007
1901	4·9	4·894	1·001	3·088	1·002	0·999
1902	4·9	5·057	0·969	3·006	0·976	0·993
1903	5·0	5·220	0·958	2·928	0·952	1·006
1904	5·0	5·383	0·929	2·856	0·978	0·950
1905	5·8	5·546	1·046	2·933	1·035	1·011
1906	6·5	5·709	1·139	3·114	1·097	1·038
1907	6·5	5·872	1·107	3·292	1·041	1·063
1908	5·3	6·035	0·878	3·124	0·979	0·897
1909	5·9	6·198	0·952	2·937	0·945	1·007
1910	6·4	6·361	1·006	2·836	0·985	1·021
1911	6·5	6·524	0·996	2·954	1·006	0·990
1912	6·8	6·687	1·017	3·019		

earlier, but once the detrended figures (which in this case equal the product of the cyclical component and the residual variations) have been found a decision has to be taken about the isolation of the cyclical component. Some form of smoothing would seem appropriate. The problem is to decide what length of moving average to use or whether in fact to bother at all. The more years that are included in a moving average, the more will the number and magnitude of cycles be eliminated, so that the statistical technique will be determining the nature of the cyclical pattern. On the other hand, if no smoothing takes place at all, on the assumption that the cyclical influence is predominant, it may prove difficult both to establish turning points and also to distinguish cycles from random and temporary disturbances in the series. A compromise which is suitable for most data is to use a three-point smoothing. The satisfactory results which this procedure produces are evident from fig. 68, where the cyclical ratio weaves a rhythmic pattern through the minor disturbances exhibited by the detrended series.

Once the cyclical component (in ratio form) has been established, the last stage in the full analysis of this data is straightforward: A/T is divided by C to give R.

6.6.2   It is seen that between 1880 and 1912 there were four major cycles in the output of steel in the United Kingdom, all of which were associated with changing commodity prices (the so-called 'Great Depression' occurred within this period); the most important of these four cycles was that which culminated in the peak of 1889. By and large the recoveries and booms were parallel to the price rises, and the slumps followed the fall in prices. Profit expectation seems to have been the key, investment slumping when expectations were poor and vice versa. The first industries to feel the impact of these cyclical tendencies were the staple and heavy industries such as steel, as a result of which the peaks and troughs in this analysis tend to be a year or so earlier than in the economy as a whole.

6.7   The correlation of time series

6.7.1   Directly connected with the analysis and isolation of the components of a time series is the interpretation of the coefficient of linear correlation calculated from two sets of chronological data. In chapter 5 we implicitly assumed that no problem existed. Now the problem and its solution must be stated. Fundamentally we know that an annual time series is made up of a trend and the residual component, while monthly or quarterly data possess in addition a seasonal component; both may also show a cyclical component. Our interest will lie in ascertaining whether the fluctuations in the two series move together or in opposite directions, or if they are not associated at all. We shall not require to know whether the trends are correlated, because by definition two linear trends must yield a unitary coefficient. However, annual data exhibit both a trend and the residual fluctuations, so that the coefficient of correlation applied to these original series will reflect the joint relationship between the trends and the fluctuations. Similarly, for data showing a seasonal component the association will be measured simultaneously between the trends, the seasonal components, and the residual fluctuations.

6.7.2   The reader may wonder why this point is being laboured. The answer is simple yet important. Suppose one is dealing with two series having trends which move in the same direction. If the association between the residual fluctuations is positive, the effect of the perfect correlation between the trends will be to increase the overall correlation above that which might be expected. If on the other hand the residual fluctuations show a high negative correlation, then the positive relationship between the trends may bring the overall correlation close to zero.

When the trends are in opposite directions, a positive association between the fluctuations will be reduced while a negative correlation will be increased.

6.7.3   To achieve meaningful results it will be necessary to correlate not the original series but, in the case mentioned above, the detrended series. By similar reasoning detrended and deseasonalized figures should be correlated in the presence of a seasonal component, while detrended, deseasonalized and decycled series should be used in cases where the cyclical influence is given specific consideration. To summarize these statements, it is evident that whatever the model the final data which should be used in the calculation of the coefficient of correlation is the residual component, R.

To illustrate the principles which have been outlined we shall consider the following data, which show the average weekly consumption of coal and oil in British industry for the years 1961 to 1965 inclusive. The two series are plotted chronologically in fig. 69(a) and (b) and a scatter diagram of the two series together is shown in fig. 69(c). From the first two it is evident that the trend of coal consumption has been falling while that of oil has risen during these years. Both series possess a marked seasonal pattern which largely coincides. The scatter diagram built up from the original data would seem to indicate little association, and this is confirmed by the calculation of r between coal and oil consumption.

Coal and oil consumption by United Kingdom industry (weekly average in thousands of tons)

Year	Quarter	Coal	Oil	Time units
1961	1	650·7	212·4	−19
	2	541·5	184·7	−17
	3	459·0	157·8	−15
	4	592·3	216·9	−13
1962	1	608·9	243·6	−11
	2	512·0	206·2	−9
	3	432·3	174·4	−7
	4	555·7	235·0	−5
1963	1	586·9	281·7	−3
	2	476·2	227·5	−1
	3	392·0	196·0	1
	4	528·4	266·6	3
1964	1	543·3	303·6	5
	2	461·9	260·0	7
	3	385·7	217·4	9
	4	512·3	300·2	11
1965	1	535·2	336·0	13
	2	442·1	281·7	15
	3	382·6	239·1	17
	4	493·4	317·3	19
Sums		10092·4	4858·1	0
Sums of squares		5202270·68	1226531·51	2660

Sums of products   2465463·30   7917·9

−8999·6

$$r = \frac{(20)(2465463\cdot30)-(10092\cdot4)(4858\cdot1)}{\sqrt{[\{(20)(5202270\cdot68)-(10092\cdot4)^2\}\ \{(20)(1226531\cdot51)-(4858\cdot1)^2\}]}}$$
$$= +0\cdot196$$

In order to obtain the detrended and deseasonalized series we must calculate the linear trends.

(a)

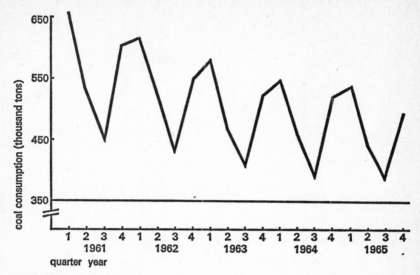

Fig. 69(a)   Time series of coal consumption.

(b)

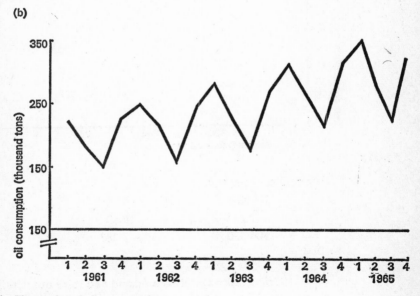

Fig. 69(b)   Time series of oil consumption.

235   The analysis of time series

(c)

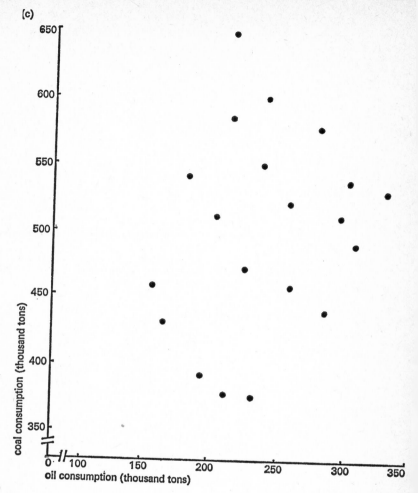

Fig. 69(c)

For coal consumption:

$$a = \frac{10092 \cdot 4}{20} \qquad b = \frac{-8999 \cdot 6}{2660}$$

$$= 504 \cdot 620 \qquad = -3 \cdot 383$$

so that

$$y_t = 504 \cdot 620 - 3 \cdot 383 x_t$$

(Origin halfway between second and third quarters of 1963; unit half quarters.)

For oil consumption:

$$a = \frac{4858 \cdot 1}{20} = 242 \cdot 905 \qquad\qquad b = \frac{7917 \cdot 9}{2660} = 2 \cdot 977$$

so that

$$y_t = 242 \cdot 905 + 2 \cdot 977 x_t$$

(Origin half-way between second and third quarters of 1963; unit half quarters.)

6.7.4   Using the multiplicative model as before, the analyses of the time series are as follows:

Coal consumption (weekly average in thousands of tons)

Seasonal analysis using multiplicative model

Year	Quarter	Actual data A	Linear trend T	Ratio detrended series $A/T = S \times R$	Seasonal index S	Residual index $A/(T \times S) = R$
1961	1	650·7	568·897	1·144	1·137	1·006
	2	541·5	562·131	0·963	0·958	1·005
	3	459·0	555·365	0·826	0·819	1·009
	4	592·3	548·599	1·080	1·086	0·994
1962	1	608·9	541·833	1·124	1·137	0·989
	2	512·0	535·067	0·957	0·958	0·999
	3	432·3	528·301	0·818	0·819	0·999
	4	555·7	521·535	1·066	1·086	0·981
1963	1	586·9	514·769	1·140	1·137	1·003
	2	476·2	508·003	0·937	0·958	0·978
	3	392·0	501·237	0·782	0·819	0·955
	4	528·4	494·471	1·069	1·086	0·984
1964	1	543·3	487·705	1·114	1·137	0·980
	2	461·9	480·939	0·960	0·958	1·002
	3	385·7	474·173	0·813	0·819	0·993
	4	512·3	467·407	1·096	1·086	1·009
1965	1	535·2	460·641	1·162	1·137	1·022
	2	442·1	453·875	0·974	0·958	1·017
	3	382·6	447·109	0·856	0·819	1·045
	4	493·4	440·343	1·120	1·086	1·031

Ratio detrended series

Year	First quarter	Second quarter	Third quarter	Fourth quarter
1961	1·144	0·963	0·826	1·080
1962	1·124	0·957	0·818	1·066
1963	1·140	0·937	0·782	1·069
1964	1·114	0·960	0·813	1·096
1965	1·162	0·974	0·856	1·120
Total	5·684	4·791	4·095	5·431
Corrected (×20/20·001)	5·684	4·791	4·095	5·431
Average (÷5)	1·137	0·958	0·819	1·086

Oil consumption (weekly average in thousands of tons)

Seasonal analysis using multiplicative model

Year	Quarter	Actual data A	Linear trend T	Ratio detrended series $A/T = S \times R$	Seasonal index S	Residual index $A/(T \times S) = R$
1961	1	212·4	186·342	1·140	1·174	0·971
	2	184·7	192·296	0·960	0·966	0·994
	3	157·8	198·250	0·796	0·800	0·995
	4	216·9	204·204	1·062	1·060	1·002
1962	1	243·6	210·158	1·159	1·174	0·987
	2	206·2	216·112	0·954	0·966	0·988
	3	174·4	222·066	0·785	0·800	0·981
	4	235·0	228·020	1·031	1·060	0·973
1963	1	281·7	233·974	1·204	1·174	1·026
	2	227·5	239·928	0·948	0·966	0·981
	3	196·0	245·882	0·797	0·800	0·996
	4	266·6	251·836	1·059	1·060	0·999
1964	1	303·6	257·790	1·178	1·174	1·003
	2	260·0	263·744	0·986	0·966	1·021
	3	217·4	269·698	0·806	0·800	1·008
	4	300·2	275·652	1·089	1·060	1·027
1965	1	336·0	281·606	1·193	1·174	1·016
	2	281·7	287·560	0·980	0·966	1·014
	3	239·1	293·514	0·815	0·800	1·019
	4	317·3	299·468	1·060	1·060	1·000

Ratio detrended series

Year	First quarter	Second quarter	Third quarter	Fourth quarter
1961	1·140	0·960	0·796	1·062
1962	1·159	0·954	0·785	1·031
1963	1·204	0·948	0·797	1·059
1964	1·178	0·986	0·806	1·089
1965	1·193	0·980	0·815	1·060
Total	5·874	4·828	3·999	5·301
Corrected $\left(\times\dfrac{20}{20\cdot002}\right)$	5·873	4·828	3·999	5·300
Average ($\div$5)	1·174	0·966	0·800	1·060

We shall now correlate the detrended series and finally the residual components, R. Denoting coal consumption by x and oil consumption by y we have the following sums, sums of squares and sums of products.

Detrended series

$$\Sigma x = 20\cdot001 \qquad \Sigma y = 20\cdot002$$
$$\Sigma x^2 = 20\cdot312177 \qquad \Sigma y^2 = 20\cdot387164$$
$$\Sigma xy = 20\cdot339427$$

Therefore

$$r = \frac{(20)(20\cdot339427)-(20\cdot001)(20\cdot002)}{\sqrt{[\{(20)(20\cdot312177)-(20\cdot001)^2\}\{(20)(20\cdot387164)-(20\cdot002)^2\}]}}$$
$$= \frac{6\cdot728538}{6\cdot894885}$$
$$= +0\cdot976$$

Detrended, deseasonalized series

$$\Sigma x = 20\cdot001 \qquad \Sigma y = 20\cdot001$$
$$\Sigma x^2 = 20\cdot009729 \qquad \Sigma y^2 = 20\cdot007419$$
$$\Sigma xy = 20\cdot004732$$

Therefore

$$r = \frac{(20)(20\cdot004732)-(20\cdot001)(20\cdot001)}{\sqrt{[\{(20)(20\cdot009729)-(20\cdot001)^2\}\{(20)(20\cdot007419)-(20\cdot001)^2\}]}}$$
$$= \frac{0\cdot054639}{0\cdot129434}$$
$$= +0\cdot422$$

6.7.5   What has emerged from these calculations? The correlation of the original data which produced $r = 0.196$ was evidently the result of the negative relationship between the trends. Once the trends were eliminated by division, the correlation between the detrended series leaps up immediately to $r = 0.976$. This reflects the fact that the seasonal components are highly correlated ($r = 0.987$) and also have little real meaning. Finally the correlation of the residual component in ratio form, eliminating both the trend and the seasonal component, gives $r = 0.422$. It seems that there is some association between the fluctuations of coal and oil consumption, but this is not as low as is suggested by the correlation of the original series and nowhere near as high as is suggested by the correlation of the detrended series. In conclusion, we have shown that direct correlation of time series can produce most misleading results. In this example, the perfect negative relationship between the trends has outweighed the high positive association in the seasonal components to reduce the true correlation between the residual fluctuations from 0.422 to 0.196.

6.8   Examples

6.8.1   *Trends*

(a) Compare from the data below the trends in volume of consumption of (i) tobacco, (ii) durable household goods, (iii) clothing, over the period 1954 to 1963. Make rough linear estimates, state the trend equations, and graph the original series and the trend estimates.

1958 constant prices (£ million)

Tobacco	Durable household goods	Clothing	Year
855	837	1,205	1954
880	934	1,297	1955
935	884	1,378	1956
981	1,005	1,439	1957
1,031	1,175	1,454	1958
1,061	1,379	1,516	1959
1,140	1,420	1,647	1960
1,213	1,388	1,709	1961
1,242	1,459	1,745	1962
1,286	1,659	1,833	1963
1,344	1,855	1,919	1964

(*National Income and Expenditure* 1965)

(b) From the following monthly data of gas production, calculate (i) a moving average estimate of the trend, and (ii) a linear trend. Plot the original series and these two estimates.

Gas available (thousands of millions of cubic feet)

	1961	*1962*	*1963*	*1964*
January		14·66	17·88	16·38
February		14·32	17·16	16·30
March		14·86	14·78	15·20
April		12·45	12·57	13·98
May		11·29	11·61	11·11
June		9·80	9·66	10·31
July	9·31	9·29	9·40	
August	8·42	8·57	8·46	
September	9·57	10·05	10·16	
October	11·13	11·15	11·62	
November	13·05	13·72	13·12	
December	14·26	14·91	15·66	

(c) Calculate a linear trend for the following data showing index numbers of industrial production (1958 = 100) per man-year for the United Kingdom.

Industrial production per man-year (1958 = 100)

1946	80	1952	89	1958	100
1947	78	1953	92	1959	105
1948	81	1954	95	1960	109
1949	85	1955	98	1961	109
1950	88	1956	97	1962	111
1951	89	1957	99		

6.8.2   *Seasonal and cyclical analysis*

(a) Using the model $A = T+S+R$ calculate a seasonally adjusted series of road deaths in Great Britain from the following monthly series:

Road deaths in Great Britain

	1961	1962	1963	1964
January		547	356	578
February		450	356	551
March		506	489	582
April		472	529	509
May		476	516	619
June		548	595	619
July	607	573	593	606
August	543	587	646	718
September	612	601	617	669
October	672	642	702	762
November	613	632	757	759
December	674	675	766	848

Produce graphs of (i) the original data, the trend and the seasonally adjusted series, and (ii) the seasonal component and detrended series. What do these suggest to you about the appropriateness of this model?

(b) Use the data of gas production in 6.8.1(b) to carry out a full time series analysis using the model

$$A = T \times S \times R$$

(c) The following data show the index of industrial production (average 1958 = 100) for all industries and the corresponding seasonally adjusted series.

Index of industrial production

Year	Quarter	Unadjusted	Seasonally adjusted
1960	1	115	112
	2	113	112
	3	106	113
	4	116	113
1961	1	116	114
	2	116	114
	3	107	114
	4	116	113

Year	Quarter	Unadjusted	Seasonally adjusted
1962	1	116	114
	2	117	115
	3	109	116
	4	118	115
1963	1	115	113
	2	119	118
	3	114	120
	4	127	124
1964	1	128	127
	2	130	127
	3	121	127
	4	134	131

(Adapted from the *Monthly Digest of Statistics*)

Calculate a moving average estimate of the trend, and by comparing the resulting values with the seasonally adjusted series estimate the residual fluctuations.

(d)

Shipbuilding: vessels commenced (thousands of tons gross)

1920	2,397	1930	950
1921	569	1931	200
1922	404	1932	72
1923	953	1933	242
1924	1,050	1934	520
1925	814	1935	683
1926	582	1936	1,081
1927	1,764	1937	1,057
1928	1,297	1938	505
1929	1,650		

From the data above calculate a linear trend and analyse the cyclical component from the model $A = T \times C \times R$. Produce diagrams showing (i) the original series and the trend, and (ii) the cyclical component and detrended series.

6.8.3 Correlation of time series

(a) The following figures relate to the production of passenger cars and the deliveries of motor-cycles by manufacturers in the United Kingdom.

Year	Production of passenger cars (thousands)	Deliveries of motor-cycles (thousands)
1954	769·2	179·6
1955	897·6	177·2
1956	707·6	124·5
1957	860·8	173·0
1958	1051·6	139·7
1959	1189·9	248·9
1960	1352·7	203·2
1961	1004·0	145·7
1962	1249·4	106·0
1963	1607·9	107·1
1964	1867·6	111·6

(*Annual Abstract of Statistics* 1965)

Calculate the coefficient of linear correlation for these data before and after the elimination of the trends. Explain carefully the meaning of your results.

(b)

Manufacturing industry

	Gross trading profits (£ million)	Gross fixed capital formation	
1954	1,763	689	1955
1955	1,936	854	1956
1956	1,934	947	1957
1957	2,059	922	1958
1958	2,006	867	1959
1959	2,245	1,021	1960
1960	2,488	1,239	1961
1961	2,343	1,168	1962
1962	2,354	1,074	1963
1963	2,535	1,244	1964

(*National Income and Expenditure* 1964)

Correlate gross trading profits in year t with gross fixed capital formation in year $t-1$. Suggest the meaning to be attached to your results.

Suggestions for further reading

For further development of general topics A. L. Edwards (*Statistical Methods for the Behavioural Sciences*; Holt, Rinehart & Winston 1964) is recommended, particularly with regard to the basic rules and principles of mathematics. The presentation of statistics is covered in detail by two works: W. J. Reichmann (*Use and Abuse of Statistics*; Methuen 1961, Penguin 1964), who clearly points out traps of presentation and interpretation, and M. E. Spear (*Charting Statistics*; McGraw-Hill 1952), which gives a comprehensive list of examples and has excellent plates that make it a standard reference on methods of conveying numerical data visually.

A. R. Ilersic (*Statistics*; H.F.L. Publishers, 13th edn 1964) deals with measures of average and dispersion, a slightly more rigorous and mathematical approach to this topic being provided by both J. Mounsey (*An Introduction to Statistical Calculations*; English Universities Press 1952) and B. C. Brookes & W. F. L. Dick (*Introduction to Statistical Methods*; Heinemann 1951); useful illustrative examples are to be found in M. R. Spiegel (*Theory and Problems of Statistics*; Schaum 1961). A very good introductory chapter on index numbers is contained in J. R. Hicks (*The Social Framework*; Oxford University Press, 3rd edn 1960), further material on this subject being available in both P. H. Karmel (*Applied Statistics for Economists*; Pitman, 2nd edn 1963) and R. G. D. Allen (*Statistics for Economists*; Hutchinson, rev. edn 1960); a good account of time series analysis is also to be found in Allen's volume whilst F. E. Croxton & D. J. Cowden (*Applied General Statistics*; Prentice-Hall, 2nd edn 1955; Pitman 1962) cover the field of changing as opposed to constant seasonal analysis.

W. Z. Hirsch (*Introduction to Modern Statistics*; Macmillan 1957) should be read in its entirety, for the author has combined humour and wit with a deep insight into his subject and the pages are leavened by a wealth of cartoons and anecdotes. G. U. Yule and M. G. Kendal (*An Introduction to the Theory of Statistics*; Griffen, 14th edn 1958) is a classic in its field, providing a detailed and thorough exposition of the entire subject. Finally reference should be made to the various government publications, such as the *Board of Trade Journal* and *Economic Trends*, from which much useful information can be obtained.

2 Logarithms

	0	1	2	3	4	5	6	7	8	9	1	2	3	4	5	6	7	8	9
10	0000	0043	0086	0128	0170	0212	0253	0294	0334	0374	4	8	13	17	21	25	29	33	37
11	0414	0453	0492	0531	0569	0607	0645	0682	0719	0755	4	8	11	15	19	23	27	30	34
12	0792	0828	0864	0899	0934	0969	1004	1038	1072	1106	4	7	10	14	17	21	24	28	31
13	1139	1173	1206	1239	1271	1303	1335	1367	1399	1430	3	6	10	13	16	19	23	26	29
14	1461	1492	1523	1553	1584	1614	1644	1673	1703	1732	3	6	9	12	15	18	21	24	27
15	1761	1790	1818	1847	1875	1903	1931	1959	1987	2014	3	6	8	11	14	17	20	23	25
16	2041	2068	2095	2122	2148	2175	2201	2227	2253	2279	3	5	8	11	13	16	18	21	24
17	2304	2330	2355	2380	2405	2430	2455	2480	2504	2529	2	5	7	10	12	15	17	20	22
18	2553	2577	2601	2625	2648	2672	2695	2718	2742	2765	2	5	7	9	12	14	16	19	21
19	2788	2810	2833	2856	2878	2900	2923	2945	2967	2989	2	4	7	9	11	13	16	18	20
20	3010	3032	3054	3075	3096	3118	3139	3160	3181	3201	2	4	6	9	11	13	15	17	19
21	3222	3243	3263	3284	3304	3324	3345	3365	3385	3404	2	4	6	8	10	12	14	16	18
22	3424	3444	3464	3483	3502	3522	3541	3560	3579	3598	2	4	6	8	10	12	14	15	17
23	3617	3636	3655	3674	3692	3711	3729	3747	3766	3784	2	4	6	7	9	11	13	15	17
24	3802	3820	3838	3856	3874	3892	3909	3927	3945	3962	2	4	5	7	9	11	12	14	16
25	3979	3997	4014	4031	4048	4065	4082	4099	4116	4133	2	3	5	7	9	10	12	14	15
26	4150	4166	4183	4200	4216	4232	4249	4265	4281	4298	2	3	5	7	8	10	12	13	15
27	4314	4330	4346	4362	4378	4393	4409	4425	4440	4456	2	3	5	6	8	9	11	13	14
28	4472	4487	4502	4518	4533	4548	4564	4579	4594	4609	2	3	5	6	8	9	11	12	14
29	4624	4639	4654	4669	4683	4698	4713	4728	4742	4757	1	3	4	6	7	9	10	12	13
30	4771	4786	4800	4814	4829	4843	4857	4871	4886	4900	1	3	4	6	7	9	10	11	13
31	4914	4928	4942	4955	4969	4983	4997	5011	5024	5038	1	3	4	6	7	8	10	11	12
32	5052	5065	5079	5092	5105	5119	5132	5145	5159	5172	1	3	4	5	7	8	9	11	12
33	5185	5198	5211	5224	5237	5250	5263	5276	5289	5302	1	3	4	5	6	8	9	10	12
34	5315	5328	5340	5353	5366	5378	5391	5403	5416	5428	1	3	4	5	6	8	9	10	11
35	5441	5453	5465	5478	5490	5502	5514	5527	5539	5551	1	2	4	5	6	7	9	10	11
36	5563	5575	5587	5599	5611	5623	5635	5647	5658	5670	1	2	4	5	6	7	8	10	11
37	5682	5694	5705	5717	5729	5740	5752	5763	5775	5786	1	2	3	5	6	7	8	9	10
38	5798	5809	5821	5832	5843	5855	5866	5877	5888	5899	1	2	3	5	6	7	8	9	10
39	5911	5922	5933	5944	5955	5966	5977	5988	5999	6010	1	2	3	4	5	7	8	9	10
40	6021	6031	6042	6053	6064	6075	6085	6096	6107	6117	1	2	3	4	5	6	8	9	10
41	6128	6138	6149	6159	6170	6180	6191	6201	6212	6222	1	2	3	4	5	6	7	8	9
42	6232	6243	6253	6263	6274	6284	6294	6304	6314	6325	1	2	3	4	5	6	7	8	9
43	6335	6345	6355	6365	6375	6385	6395	6405	6415	6425	1	2	3	4	5	6	7	8	9
44	6435	6444	6454	6464	6474	6484	6493	6503	6513	6522	1	2	3	4	5	6	7	8	9
45	6532	6542	6551	6561	6571	6580	6590	6599	6609	6618	1	2	3	4	5	6	7	8	9
46	6628	6637	6646	6656	6665	6675	6684	6693	6702	6712	1	2	3	4	5	6	7	7	8
47	6721	6730	6739	6749	6758	6767	6776	6785	6794	6803	1	2	3	4	5	5	6	7	8
48	6812	6821	6830	6839	6848	6857	6866	6875	6884	6893	1	2	3	4	4	5	6	7	8
49	6902	6911	6920	6928	6937	6946	6955	6964	6972	6981	1	2	3	4	4	5	6	7	8
50	6990	6998	7007	7016	7024	7033	7042	7050	7059	7067	1	2	3	3	4	5	6	7	8
51	7076	7084	7093	7101	7110	7118	7126	7135	7143	7152	1	2	3	3	4	5	6	7	8
52	7160	7168	7177	7185	7193	7202	7210	7218	7226	7235	1	2	2	3	4	5	6	7	7
53	7243	7251	7259	7267	7275	7284	7292	7300	7308	7316	1	2	2	3	4	5	6	7	7
54	7324	7332	7340	7348	7356	7364	7372	7380	7388	7396	1	2	2	3	4	5	6	6	7

	0	1	2	3	4	5	6	7	8	9	Mean differences 1	2	3	4	5	6	7	8	9
55	7404	7412	7419	7427	7435	7443	7451	7459	7466	7474	1	2	2	3	4	5	5	6	7
56	7482	7490	7497	7505	7513	7520	7528	7536	7543	7551	1	2	2	3	4	5	5	6	7
57	7559	7566	7574	7582	7589	7597	7604	7612	7619	7627	1	2	2	3	4	5	5	6	7
58	7634	7642	7649	7657	7664	7672	7679	7686	7694	7701	1	1	2	3	4	4	5	6	7
59	7709	7716	7723	7731	7738	7745	7752	7760	7767	7774	1	1	2	3	4	4	5	6	7
60	7782	7789	7796	7803	7810	7818	7825	7832	7839	7846	1	1	2	3	4	4	5	6	6
61	7853	7860	7868	7875	7882	7889	7896	7903	7910	7917	1	1	2	3	4	4	5	6	6
62	7924	7931	7938	7945	7952	7959	7966	7973	7980	7987	1	1	2	3	3	4	5	6	6
63	7993	8000	8007	8014	8021	8028	8035	8041	8048	8055	1	1	2	3	3	4	5	5	6
64	8062	8069	8075	8082	8089	8096	8102	8109	8116	8122	1	1	2	3	3	4	5	5	6
65	8129	8136	8142	8149	8156	8162	8169	8176	8182	8189	1	1	2	3	3	4	5	5	6
66	8195	8202	8209	8215	8222	8228	8235	8241	8248	8254	1	1	2	3	3	4	5	5	6
67	8261	8267	8274	8280	8287	8293	8299	8306	8312	8319	1	1	2	3	3	4	5	5	6
68	8325	8331	8338	8344	8351	8357	8363	8370	8376	8382	1	1	2	3	3	4	4	5	6
69	8388	8395	8401	8407	8414	8420	8426	8432	8439	8445	1	1	2	3	3	4	4	5	6
70	8451	8457	8463	8470	8476	8482	8488	8494	8500	8506	1	1	2	2	3	4	4	5	6
71	8513	8519	8525	8531	8537	8543	8549	8555	8561	8567	1	1	2	2	3	4	4	5	5
72	8573	8579	8585	8591	8597	8603	8609	8615	8621	8627	1	1	2	2	3	4	4	5	5
73	8633	8639	8645	8651	8657	8663	8669	8675	8681	8686	1	1	2	2	3	4	4	5	5
74	8692	8698	8704	8710	8716	8722	8727	8733	8739	8745	1	1	2	2	3	4	4	5	5
75	8751	8756	8762	8768	8774	8779	8785	8791	8797	8802	1	1	2	2	3	3	4	5	5
76	8808	8814	8820	8825	8831	8837	8842	8848	8854	8859	1	1	2	2	3	3	4	5	5
77	8865	8871	8876	8882	8887	8893	8899	8904	8910	8915	1	1	2	2	3	3	4	4	5
78	8921	8927	8932	8938	8943	8949	8954	8960	8965	8971	1	1	2	2	3	3	4	4	5
79	8976	8982	8987	8993	8998	9004	9009	9015	9020	9025	1	1	2	2	3	3	4	4	5
80	9031	9036	9042	9047	9053	9058	9063	9069	9074	9079	1	1	2	2	3	3	4	4	5
81	9085	9090	9096	9101	9106	9112	9117	9122	9128	9133	1	1	2	2	3	3	4	4	5
82	9138	9143	9149	9154	9159	9165	9170	9175	9180	9186	1	1	2	2	3	3	4	4	5
83	9191	9196	9201	9206	9212	9217	9222	9227	9232	9238	1	1	2	2	3	3	4	4	5
84	9243	9248	9253	9258	9263	9269	9274	9279	9284	9289	1	1	2	2	3	3	4	4	5
85	9294	9299	9304	9309	9315	9320	9325	9330	9335	9340	1	1	2	2	3	3	4	4	5
86	9345	9350	9355	9360	9365	9370	9375	9380	9385	9390	1	1	2	2	3	3	4	4	5
87	9395	9400	9405	9410	9415	9420	9425	9430	9435	9440	0	1	1	2	2	3	3	4	4
88	9445	9450	9455	9460	9465	9469	9474	9479	9484	9489	0	1	1	2	2	3	3	4	4
89	9494	9499	9504	9509	9513	9518	9523	9528	9533	9538	0	1	1	2	2	3	3	4	4
90	9542	9547	9552	9557	9562	9566	9571	9576	9581	9586	0	1	1	2	2	3	3	4	4
91	9590	9595	9600	9605	9609	9614	9619	9624	9628	9633	0	1	1	2	2	3	3	4	4
92	9638	9643	9647	9652	9657	9661	9666	9671	9675	9680	0	1	1	2	2	3	3	4	4
93	9685	9689	9694	9699	9703	9708	9713	9717	9722	9727	0	1	1	2	2	3	3	4	4
94	9731	9736	9741	9745	9750	9754	9759	9763	9768	9773	0	1	1	2	2	3	3	4	4
95	9777	9782	9786	9791	9795	9800	9805	9809	9814	9818	0	1	1	2	2	3	3	4	4
96	9823	9827	9832	9836	9841	9845	9850	9854	9859	9863	0	1	1	2	2	3	3	4	4
97	9868	9872	9877	9881	9886	9890	9894	9899	9903	9908	0	1	1	2	2	3	3	4	4
98	9912	9917	9921	9926	9930	9934	9939	9943	9948	9952	0	1	1	2	2	3	3	4	4
99	9956	9961	9965	9969	9974	9978	9983	9987	9991	9996	0	1	1	2	2	3	3	3	4

3 Antilogarithms

	0	1	2	3	4	5	6	7	8	9	1	2	3	4	5	6	7	8	9
·00	1000	1002	1005	1007	1009	1012	1014	1016	1019	1021	0	0	1	1	1	1	2	2	2
·01	1023	1026	1028	1030	1033	1035	1038	1040	1042	1045	0	0	1	1	1	1	2	2	2
·02	1047	1050	1052	1054	1057	1059	1062	1064	1067	1069	0	0	1	1	1	1	2	2	2
·03	1072	1074	1076	1079	1081	1084	1086	1089	1091	1094	0	0	1	1	1	1	2	2	2
·04	1096	1099	1102	1104	1107	1109	1112	1114	1117	1119	0	1	1	1	1	2	2	2	2
·05	1122	1125	1127	1130	1132	1135	1138	1140	1143	1146	0	1	1	1	1	2	2	2	2
·06	1148	1151	1153	1156	1159	1161	1164	1167	1169	1172	0	1	1	1	1	2	2	2	2
·07	1175	1178	1180	1183	1186	1189	1191	1194	1197	1199	0	1	1	1	1	2	2	2	2
·08	1202	1205	1208	1211	1213	1216	1219	1222	1225	1227	0	1	1	1	1	2	2	2	3
·09	1230	1233	1236	1239	1242	1245	1247	1250	1253	1256	0	1	1	1	1	2	2	2	3
·10	1259	1262	1265	1268	1271	1274	1276	1279	1282	1285	0	1	1	1	1	2	2	2	3
·11	1288	1291	1294	1297	1300	1303	1306	1309	1312	1315	0	1	1	1	1	2	2	2	3
·12	1318	1321	1324	1327	1330	1334	1337	1340	1343	1346	0	1	1	1	2	2	2	2	3
·13	1349	1352	1355	1358	1361	1365	1368	1371	1374	1377	0	1	1	1	2	2	2	3	3
·14	1380	1384	1387	1390	1393	1396	1400	1403	1406	1409	0	1	1	1	2	2	2	3	3
·15	1413	1416	1419	1422	1426	1429	1432	1435	1439	1442	0	1	1	1	2	2	2	3	3
·16	1445	1449	1452	1455	1459	1462	1466	1469	1472	1476	0	1	1	1	2	2	2	3	3
·17	1479	1483	1486	1489	1493	1496	1500	1503	1507	1510	0	1	1	1	2	2	2	3	3
·18	1514	1517	1521	1524	1528	1531	1535	1538	1542	1545	0	1	1	1	2	2	2	3	3
·19	1549	1552	1556	1560	1563	1567	1570	1574	1578	1581	0	1	1	1	2	2	3	3	3
·20	1585	1589	1592	1596	1600	1603	1607	1611	1614	1618	0	1	1	1	2	2	3	3	3
·21	1622	1626	1629	1633	1637	1641	1644	1648	1652	1656	0	1	1	2	2	2	3	3	3
·22	1660	1663	1667	1671	1675	1679	1683	1687	1690	1694	0	1	1	2	2	2	3	3	3
·23	1698	1702	1706	1710	1714	1718	1722	1726	1730	1734	0	1	1	2	2	2	3	3	4
·24	1738	1742	1746	1750	1754	1758	1762	1766	1770	1774	0	1	1	2	2	2	3	3	4
·25	1778	1782	1786	1791	1795	1799	1803	1807	1811	1816	0	1	1	2	2	2	3	3	4
·26	1820	1824	1828	1832	1837	1841	1845	1849	1854	1858	0	1	1	2	2	3	3	3	4
·27	1862	1866	1871	1875	1879	1884	1888	1892	1897	1901	0	1	1	2	2	3	3	3	4
·28	1905	1910	1914	1919	1923	1928	1932	1936	1941	1945	0	1	1	2	2	3	3	4	4
·29	1950	1954	1959	1963	1968	1972	1977	1982	1986	1991	0	1	1	2	2	3	3	4	4
·30	1995	2000	2004	2009	2014	2018	2023	2028	2032	2037	0	1	1	2	2	3	3	4	4
·31	2042	2046	2051	2056	2061	2065	2070	2075	2080	2084	0	1	1	2	2	3	3	4	4
·32	2089	2094	2099	2104	2109	2113	2118	2123	2128	2133	0	1	1	2	2	3	3	4	4
·33	2138	2143	2148	2153	2158	2163	2168	2173	2178	2183	0	1	1	2	2	3	3	4	4
·34	2188	2193	2198	2203	2208	2213	2218	2223	2228	2234	1	1	2	2	3	3	4	4	5
·35	2239	2244	2249	2254	2259	2265	2270	2275	2280	2286	1	1	2	2	3	3	4	4	5
·36	2291	2296	2301	2307	2312	2317	2323	2328	2333	2339	1	1	2	2	3	3	4	4	5
·37	2344	2350	2355	2360	2366	2371	2377	2382	2388	2393	1	1	2	2	3	3	4	4	5
·38	2399	2404	2410	2415	2421	2427	2432	2438	2443	2449	1	1	2	2	3	3	4	4	5
·39	2455	2460	2466	2472	2477	2483	2489	2495	2500	2506	1	1	2	2	3	3	4	5	5
·40	2512	2518	2523	2529	2535	2541	2547	2553	2559	2564	1	1	2	2	3	4	4	5	5
·41	2570	2576	2582	2588	2594	2600	2606	2612	2618	2624	1	1	2	2	3	4	4	5	5
·42	2630	2636	2642	2648	2655	2661	2667	2673	2679	2685	1	1	2	2	3	4	4	5	6
·43	2692	2698	2704	2710	2716	2723	2729	2735	2742	2748	1	1	2	3	3	4	4	5	6
·44	2754	2761	2767	2773	2780	2786	2793	2799	2805	2812	1	1	2	3	3	4	4	5	6
·45	2818	2825	2831	2838	2844	2851	2858	2864	2871	2877	1	1	2	3	3	4	5	5	6
·46	2884	2891	2897	2904	2911	2917	2924	2931	2938	2944	1	1	2	3	3	4	5	5	6
·47	2951	2958	2965	2972	2979	2985	2992	2999	3006	3013	1	1	2	3	3	4	5	5	6
·48	3020	3027	3034	3041	3048	3055	3062	3069	3076	3083	1	1	2	3	4	4	5	6	6
·49	3090	3097	3105	3112	3119	3126	3133	3141	3148	3155	1	1	2	3	4	4	5	6	6

	0	1	2	3	4	5	6	7	8	9	1	2	3	4	5	6	7	8	9
·50	3162	3170	3177	3184	3192	3199	3206	3214	3221	3228	1	1	2	3	4	4	5	6	7
·51	3236	3243	3251	3258	3266	3273	3281	3289	3296	3304	1	2	2	3	4	5	5	6	7
·52	3311	3319	3327	3334	3342	3350	3357	3365	3373	3381	1	2	2	3	4	5	5	6	7
·53	3388	3396	3404	3412	3420	3428	3436	3443	3451	3459	1	2	2	3	4	5	6	6	7
·54	3467	3475	3483	3491	3499	3508	3516	3524	3532	3540	1	2	2	3	4	5	6	6	7
·55	3548	3556	3565	3573	3581	3589	3597	3606	3614	3622	1	2	2	3	4	5	6	7	7
·56	3631	3639	3648	3656	3664	3673	3681	3690	3698	3707	1	2	3	3	4	5	6	7	8
·57	3715	3724	3733	3741	3750	3758	3767	3776	3784	3793	1	2	3	3	4	5	6	7	8
·58	3802	3811	3819	3828	3837	3846	3855	3864	3873	3882	1	2	3	4	4	5	6	7	8
·59	3890	3899	3908	3917	3926	3936	3945	3954	3963	3972	1	2	3	4	5	5	6	7	8
·60	3981	3990	3999	4009	4018	4027	4036	4046	4055	4064	1	2	3	4	5	6	6	7	8
·61	4074	4083	4093	4102	4111	4121	4130	4140	4150	4159	1	2	3	4	5	6	7	8	9
·62	4169	4178	4188	4198	4207	4217	4227	4236	4246	4256	1	2	3	4	5	6	7	8	9
·63	4266	4276	4285	4295	4305	4315	4325	4335	4345	4355	1	2	3	4	5	6	7	8	9
·64	4365	4375	4385	4395	4406	4416	4426	4436	4446	4457	1	2	3	4	5	6	7	8	9
·65	4467	4477	4487	4498	4508	4519	4529	4539	4550	4560	1	2	3	4	5	6	7	8	9
·66	4571	4581	4592	4603	4613	4624	4634	4645	4656	4667	1	2	3	4	5	6	7	9	10
·67	4677	4688	4699	4710	4721	4732	4742	4753	4764	4775	1	2	3	4	5	7	8	9	10
·68	4786	4797	4808	4819	4831	4842	4853	4864	4875	4887	1	2	3	4	6	7	8	9	10
·69	4898	4909	4920	4932	4943	4955	4966	4977	4989	5000	1	2	3	5	6	7	8	9	10
·70	5012	5023	5035	5047	5058	5070	5082	5093	5105	5117	1	2	3	5	6	7	8	9	10
·71	5129	5140	5152	5164	5176	5188	5200	5212	5224	5236	1	2	4	5	6	7	8	10	11
·72	5248	5260	5272	5284	5297	5309	5321	5333	5346	5358	1	2	4	5	6	7	9	10	11
·73	5370	5383	5395	5408	5420	5433	5445	5458	5470	5483	1	2	4	5	6	7	9	10	11
·74	5495	5508	5521	5534	5546	5559	5572	5585	5598	5610	1	3	4	5	6	8	9	10	12
·75	5623	5636	5649	5662	5675	5689	5702	5715	5728	5741	1	3	4	5	7	8	9	10	12
·76	5754	5768	5781	5794	5808	5821	5834	5848	5861	5875	1	3	4	5	7	8	9	11	12
·77	5888	5902	5916	5929	5943	5957	5970	5984	5998	6012	1	3	4	5	7	8	10	11	12
·78	6026	6039	6053	6067	6081	6095	6109	6124	6138	6152	1	3	4	6	7	8	10	11	13
·79	6166	6180	6194	6209	6223	6237	6252	6266	6281	6295	1	3	4	6	7	9	10	11	13
·80	6310	6324	6339	6353	6368	6383	6397	6412	6427	6442	1	3	4	6	7	9	10	12	13
·81	6457	6471	6486	6501	6516	6531	6546	6561	6577	6592	2	3	5	6	8	9	11	12	14
·82	6607	6622	6637	6653	6668	6683	6699	6714	6730	6745	2	3	5	6	8	9	11	12	14
·83	6761	6776	6792	6808	6823	6839	6855	6871	6887	6902	2	3	5	6	8	9	11	13	14
·84	6918	6934	6950	6966	6982	6998	7015	7031	7047	7063	2	3	5	6	8	10	11	13	14
·85	7079	7096	7112	7129	7145	7161	7178	7194	7211	7228	2	3	5	7	8	10	12	13	15
·86	7244	7261	7278	7295	7311	7328	7345	7362	7379	7396	2	3	5	7	8	10	12	13	15
·87	7413	7430	7447	7464	7482	7499	7516	7534	7551	7568	2	3	5	7	9	10	12	14	16
·88	7586	7603	7621	7638	7656	7674	7691	7709	7727	7745	2	4	5	7	9	11	12	14	16
·89	7762	7780	7798	7816	7834	7852	7870	7889	7907	7925	2	4	5	7	9	11	13	14	16
·90	7943	7962	7980	7998	8017	8035	8054	8072	8091	8110	2	4	6	7	9	11	13	15	17
·91	8128	8147	8166	8185	8204	8222	8241	8260	8279	8299	2	4	6	8	9	11	13	15	17
·92	8318	8337	8356	8375	8395	8414	8433	8453	8472	8492	2	4	6	8	10	12	14	15	17
·93	8511	8531	8551	8570	8590	8610	8630	8650	8670	8690	2	4	6	8	10	12	14	16	18
·94	8710	8730	8750	8770	8790	8810	8831	8851	8872	8892	2	4	6	8	10	12	14	16	18
·95	8913	8933	8954	8974	8995	9016	9036	9057	9078	9099	2	4	6	8	10	12	15	17	19
·96	9120	9141	9162	9183	9204	9226	9247	9268	9290	9311	2	4	6	8	11	13	15	17	19
·97	9333	9354	9376	9397	9419	9441	9462	9484	9506	9528	2	4	7	9	11	13	15	17	20
·98	9550	9572	9594	9616	9638	9661	9683	9705	9727	9750	2	4	7	9	11	13	16	18	20
·99	9772	9795	9817	9840	9863	9886	9908	9931	9954	9977	2	5	7	9	11	14	16	18	20

Answers to Examples

Chapter 1

1.7.1
(a) $48\frac{2}{3}$ (b) 408 (c) $156\frac{1}{8}$ (d) 7 (e) -6
(f) 0 (g) $-2\frac{7}{8}$ (h) -12 (i) $-\frac{1}{4}$ (j) -14280

1.7.2
(a) $1\frac{7}{40}$ (b) $\frac{26}{35}$ (c) $3\frac{3}{5}$ (d) $1\frac{4}{15}$ (e) $6\frac{2}{3}$
(f) $0\cdot00056885$ (g) $380\cdot01289$ (h) $0\cdot05342$ (i) $-14\cdot9701$
(j) $-2\cdot9866$

1.7.3
(f) $a = 0\cdot76, b = 1\cdot62$ (g) $a = 1, b = -2\cdot5$
(h) $a = 12\cdot1, c = -0\cdot1, b = 1\cdot5$

1.7.4
(a) $17\cdot6775$ (b) $129\cdot6$ (c) $8^4 = 4096$ (d) 216

(e) $\dfrac{1}{\sqrt{33}} = 0\cdot1741$ (f) $8\cdot840$ (g) $0\cdot000848$ (h) 2159

(i) 2784 (j) 36,330,000

1.7.5
(a) 85 (b) 1,385 (c) 285 (d) 3,261 (e) 23,553 (f) 81,225
(g) 94 (h) -133 (i) $-1,431$ (j) $-2,944,614$

1.7.6

(a) 262, 224, 250, 212 $(++; +-; -+; --)$
(b) 5, 15, -1, 9 $(++; +-; -+; --)$
(c) 450, 400, 198, 176 $(++; +-; -+; --)$
(d) 1·66, 2·03, 1·50, 1·83 $(++; +-; -+; --)$
(e) 3·35, 14·5, 2·96, 12·83 $(++; +-; -+; --)$

Chapter 3

3.11

(a) $\bar{x} = 1·95$ faults, $s = 2·01$ faults (median $= 1$)
(b) $\bar{x} = £49·9$ million, $s = £176·53$ million
(c) Median $= £506·60$, quartile deviation $= £168·90$
(d) 9·43 per cent
(e) $\bar{x} = 2·64$, median $= 2$, mode $= 2$
(f) (i) $A = 7·35$ per cent; $B = 8·7$ per cent
    (ii) $A = 89·8$ per cent; $B = 58·2$ per cent
(g) 30·73 m.p.g.; 27·79 m.p.g.
(h) (i) 67·7 per cent
    (ii) 37·01 per cent
(i) Depot I $= 493$ units; Depot II $= 583$ units
(j) $\bar{x} = 36·6$ years, $s = 31·7$ years
(k) $1953 = 1·371$, $1963 = 1·375$
(l)

	Vehicles	Food, drink and tobacco
Q_1	33·3	23·5
M	78·0	63·0
Q_3	328·6	155·5
$Q.D.$	147·65	66·0

Chapter 4

4.8.1

(a) *Laspeyres*   *Paasche*
    $I_{1950} = 131·9$ $I_{1950} = 126·2$
    $I_{1960} = 212·7$ $I_{1960} = 157·7$
(b) (i) Index for 16 January   Index for 17 January
        $1962 = 100$ $1956 = 100$
        $I_{1963} = 103·67$ $I_{1963} = 121·81$
        $I_{1965} = 112·16$ $I_{1965} = 131·79$
    (ii) $I_{1963} = 124·96$
        $I_{1965} = 136·20$

(c)	Index of weekly earnings	Index of hours of work (January 1956 = 100)	Index of output
	100	100	100
	101·5	100·2	104·0
	102·3	98·6	102·5
	104·1	100·8	106·3
	104·8	98·5	106·5

(d)		Food, drink and tobacco	Housing, fuel and light
	1955	93·13	81·45
	1958	100	100
	1961	102·91	110·43
	1963	108·77	118·29

Paasche aggregative index

(e)	1962	103
	1963	104·85
	1964	106·60

Base weighted

Chapter 5

5.9.1
(a) $y = 28·344 + 0·241x$
(b) $y = 7·755 + 0·292x$
(c) £1·36$\frac{1}{2}$
(d) $y = 10·2 + 0·11x$
(e) $y = 2·308 + 0·106x$

5.9.2
(a) $r = 0·986 (r^2 = 0·972)$
(b) $r = 0·871$
(c) $r = -0·21$
(d) $r = 0·792$
(e) $r = 0·8997$

Chapter 6

6.8.1
(a) Tobacco: $y_t = 1088 + 50·573x_t$
 Durable household goods: $y_t = 1272·27 + 95·509x_t$
 Clothing: $y_t = 1558·4 + 68·618x_t$
 (For all three: origin 1959, unit one year.)

(b) (i)

	1962	1963
January	11·93	12·62
February	11·93	12·62
March	11·96	12·62
April	11·98	12·64
May	12·01	12·64
June	12·06	12·64
July	12·22	12·61
August	12·48	12·51
September	12·59	12·49
October	12·59	12·57
November	12·61	12·61
December	12·62	12·61

(ii) $y_t = 12\cdot403 + 0\cdot0283x_t$

(Origin half-way between December 1962 and January 1963, unit half month.)

(c) $y_t = 94\cdot4118 + 2\cdot0417x_t$

(Origin 1954; unit one year.)

6.8.2

(a)

	1961	1962	1963	1964
January		631	440	662
February		578	484	679
March		564	547	640
April		554	611	591
May		528	568	671
June		554	601	625
July	593	559	579	592
August	498	542	601	673
September	578	567	583	635
October	577	547	607	667
November	500	519	644	646
December	539	540	631	713

6.8.3

(a) Before −0·2963

After +0·2317 $(A = T + R)$

(b) $r = 0\cdot8235$ $(A = T \times R)$

Index

Studies in Applied Statistics

The series is designed to help the student and the worker whose own mathematics is limited but who finds himself increasingly in need of statistical techniques. *Statistics for the Social Scientist: 1 Introducing Statistics* provides a reminder of elementary work and a background to further study in various areas.

Statistics for the Social Scientist: 2 Applied Statistics
K. A. Yeomans

Includes sampling, multiple regression, an introduction to non-parametric methods, and other topics of importance to the social worker.

Contents

1 The principles of statistical sampling
2 Testing hypotheses
3 Statistical sampling practice
4 Non-linear and multiple regression and correlation
5 Mathematical trend curves and exponential smoothing
6 Non-parametric methods
7 Econometrics and operational research

Statistics for Technology
C. Chatfield

Provides a sound basis of statistical knowledge to meet the demands created by the increasing use of statistical methods in engineering and applied science. The mathematics is kept as simple as possible. The book is in three parts. The first part introduces the ideas and simple ways of data analysis, the second explains the theory, and the third part deals with applications to experimental design and analysis, quality control and life testing. It includes many topics not previously brought together in one book.

Other Penguin Education Titles on Statistics

Statistics for the Teacher
A. C. Crocker

An introduction to statistics for use by the teacher-in-training, to assist him in understanding the frequent references to statistics in his reading or practical work. It is written in a simple, attractive style and assumes very little previous mathematical knowledge whilst covering a wide range of statistical topics including correlation, significance, item analysis and the reliability of tests.

Basic Statistics in Behavioural Research
A. E. Maxwell

This book, written to provide medical students on a postgraduate course in psychiatry with some initial understanding of the statistical terminology and elementary techniques, will be useful to any student of the behavioural sciences who wants a simple introductory course on the principles of experimental design and data analysis.

The author takes account of the fact that many of his readers will have forgotten their school mathematics.

Experiment, Design and Statistics in Psychology
Colin Robson

This book aims to assist the reader to design, carry out, analyse and interpret experiments. It emphasizes the close relationship between experimental design and statistical analysis and, though confined primarily to techniques used in psychology and education, offers valuable advice to anyone interested in performing experiments. The author adopts a refreshing, essentially non-mathematical, approach and provides a detailed 'recipe', linked to a fully-worked example, for each statistical test he discusses.